AF836861

LUMINESCENCE OF BIOPOLYMERS AND CELLS

LUMINESCENCE OF BIOPOLYMERS AND CELLS

Grigorii M. Barenboim, Aleksandr N. Domanskii, and Konstantin K. Turoverov
Institute of Cytology
Academy of Sciences of the USSR
Leningrad, USSR

Translated from Russian

Translation Editor Raymond F. Chen
Laboratory of Technical Development
National Heart Institute
National Institutes of Health
Bethesda, Maryland

Springer Science+Business Media, LLC 1969

Grigorii Matveevich Barenboim is a physical chemist. He attained the degree of Candidate of Physicomathematical Sciences for his studies of the role of excited electronic states of biopolymers in radiobiological processes. About 40 studies in the field of radioluminescence and photoluminescence of biopolymers in cells have been conducted under his supervision. He is currently the supervisor of a group engaged in luminescence research at the Institute of Cytology of the Academy of Sciences of the USSR in Leningrad and gives a course of lectures on quantum biophysics at Leningrad University.

Aleksandr Nikolaevich Domanskii was graduated from Leningrad Polytechnic Institute in 1962 with a degree in isotope physics. He has been associated with work in the luminescence research group at the Institute of Cytology and has specialized in the field of radioluminescence, quantitative methods, and apparatus for luminescence analysis.

Konstantin Konstantinovich Turoverov was graduated from Leningrad State University in 1962 with a degree in polymer physics. He is now associated with the Division of Marine Biology of the Far-Eastern Branch of the Academy of Sciences after having worked with the luminescence research group of the Institute of Cytology. His work has been focused mainly on the physical mechanisms of the influence of conformational transformations of biopolymers on their luminescence.

The original Russian text, published for the Institute of Cytology of the Academy of Sciences of the USSR by Nauka Press in Leningrad in 1966, has been thoroughly revised by the authors for the present edition.

Григорий Матвеевич Баренбойм,
Александр Николаевич Доманский,
Константин Константинович Туроверов

ЛЮМИНЕСЦЕНЦИЯ БИОПОЛИМЕРОВ И КЛЕТОК
LYUMINESTSENTSIYA BIOPOLIMEROV I KLETOK

Library of Congress Catalog Card Number 68-26768

ISBN 978-1-4899-6216-4 ISBN 978-1-4899-6441-0 (eBook)
DOI 10.1007/978-1-4899-6441-0

© 1969 Springer Science+Business Media New York

Originally published by Plenum press, New York in 1969.

Softcover reprint of the hardcover 1st edition 1969

All rights reserved

No part of this publication may be reproduced in any form without written permission from the publisher

PREFACE

During the last 15 years, the specialized roles played by tryptophan residues in proteins have gradually become evident. The unique structure of tryptophan, so unlike that of any other naturally occurring amino acid, is responsible for its ability to form charge-transfer complexes, make up parts of the active-site regions of dehydrogenases and other enzymes, form sections of "hydrophobic" regions of proteins, and probably play a key role in the binding regions of antibodies and serum albumins. The indole nucleus endows tryptophan with the ability to act in these roles and is responsible for the ability to emit fluorescence in the ultraviolet and phosphorescence in the visible region. At the same time, it has become clear that tryptophan residues are by far the most important groups responsible for the luminescence of proteins and intact cells. The importance of tryptophan is one of the main reasons why fluorescence and other forms of luminescence spectroscopy are rapidly developing into powerful investigative tools of the molecular biologist.

Preeminent as it is in the study of biological macromolecules, fluorescence is but one aspect of luminescence. The work of Barenboim and his colleagues has contributed not only to fluorescence spectroscopy but also to other areas where light emission is encountered. These areas include phosphorescence, biological chemiluminescence, light emission following activation by high-energy radiation, as well as ultraviolet fluorescence microscopy of cells and subcellular particles. The breadth of interest of the authors, as evidenced by this book, seems entirely natural, since all these phenomena should be discussed together. Indeed, it is the arbitrary classification of luminescence into separate categories which is somewhat artificial.

The general reader will find the basic principles stated in straightforward terms, and the review of luminescence research is an excellent introduction to work in this field. One notes some subjects described here which are not, to any extent, being pursued outside the USSR. Of particular interest are the studies on the "ultraweak" ultraviolet chemiluminescence of cells and organelles, and the use of ultraviolet fluorescence microscopy to observe cellular structure using the intrinsic protein fluorescence. In addition, the serious worker will find the last two chapters, which deal with instrumentation, quite valuable. There are few, if any, comparable descriptions of Soviet photometric apparatus available in English.

For many years the contributions of Russian scientists to the understanding of luminescence phenomena have been most impressive. The already classical studies of Vavilov, Sveshnikov, Terenin, Feofilov, Neporent, and others are examples. Several laboratories in the USSR have carried this long tradition of luminescence research into the biological area and have been most active since the discovery of protein fluorescence in 1956. The contributions of groups headed by S. V. Konev and Yu. A. Vladimirov are numerous and well known, and the importance and originality of work by Barenboim's laboratory should also be appreciated. The present book reviews much Soviet research including that of the authors, which may not be familiar to scientists in the West.

Raymond F. Chen
National Heart Institute

Contents

INTRODUCTION

LUMINESCENCE OF SOLUTIONS OF POLYATOMIC ORGANIC MOLECULES

The physical basis of luminescence is treated in a large number of books and reviews. These include, in increasing degree of complexity, the books by Konstantinova-Shlezinger (1961), Udenfriend (1962), West (1956), Terenin (1947), Levshin (1951), Pringsheim (1949), El'yashevich (1962), Vol'kenshtein (1955), Stepanov (1955), and Stepanov and Gribkovskii (1963). This introduction is a summary of the fundamentals of molecular luminescence which the reader may find useful in reading this book. It is based in part on the monographs listed above, but also on various reviews published in the last few years, mainly those devoted to the luminescence of complex, and above all, heteroatomic molecules.

0.1. MOLECULAR SPECTRA AND BASIC CONCEPTS OF LUMINESCENCE

The energy E of a molecule consists of its kinetic energy E_{kin} the energy E_{el} associated with the motion of the electrons, the vibrational energy E_{vib} of the nuclei, and the rotational energy E_{rot} of the molecule as a whole:

$$E = E_{\text{kin}} + E_{\text{el}} + E_{\text{vib}} + E_{\text{rot}} \tag{0.1}$$

This formula is valid only in the first approximation since it does not take into account the interaction between the three forms of internal motion, which is particularly important for complex molecules. However, Eq. (0.1) enables us to construct energy diagrams in which an energy level term corresponds to each electronic state. Moreover, each electronic level has a number of vibrational sublevels, and each vibrational sublevel, a number of rotational sublevels (Fig. 0.1).

The lowest possible electronic energy (E_0 in Fig. 0.1) corresponds to the ground (unexcited) state of the molecule. Electronic states with higher energy ($E_1, E_2, \ldots, E_n$) are called excited states.

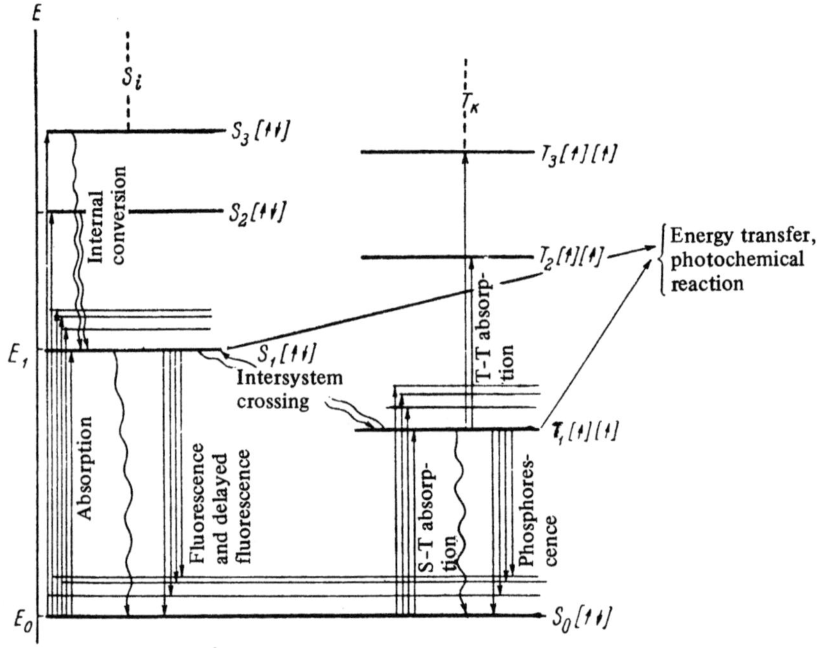

Fig. 0.1. Molecular energy levels and transitions between them. The level energy E is marked on the vertical axis. S_0 is the ground state, $S_1, S_2, \ldots, S_i$, and $T_1, T_2, \ldots, T_k$ are singlet and triplet levels, respectively. Thick horizontal lines represent electronic levels, thin horizontal lines represent vibrational sublevels. Vertical lines indicate transitions between the electronic levels accompanied by the absorption or emission of light. Wavy lines represent radiationless transitions. Spin directions of electrons in the outer orbitals of a molecule in the corresponding electronic states are shown schematically in square brackets.

If the frequency ν of a photon incident on the molecule satisfies the relation

$$h\nu = E_1 - E_0 = \Delta E \tag{0.2}$$

where $h = 6.625 \times 10^{-27}$, erg·sec (Planck's constant), the molecule can absorb the quantum $h\nu$ and undergo a transition from the ground state E_0 to a higher-lying level E_1.(One often speaks of electronic transitions in molecules because the transition of a molecule from one electronic state to another is connected with a change in the electron-cloud configuration.)

The reverse transition, i.e., the transition from E_1 to E_0 may be accompanied by the emission of a photon of energy $h\nu$, i.e., the emission of luminescence.* Depending on the type of excitation, the emission is called

*A molecule can, of course , undergo a transition to an excited state not only as a result of absorption of a photon, but also in the course of chemical reaction, or under the action of various types of radiation.

photo-, chemi-, or radioluminescence or is characterized by some other suitable prefix. In this introduction we shall consider luminescence which is excited by light.

Depending on the nature of the excited state and the nature of the transition to the ground state, the emitted luminescence may be classified as fluorescence or phosphorescence. The distinction between these two types of luminescence will be discussed in greater detail in Section 0.2. Relaxation from the excited state to the ground state occurs in a very short but measurable interval of time. For many organic molecules in solution this time is not less than 10^{-8} to 10^{-9} sec (fluorescence), which is considerably longer than the period of the light waves. This is the most characteristic feature of luminescence and distinguishes it from Rayleigh and Raman scattering, Vavilov–Cerenkov emission, refraction and reflection of light, and so on. These are not associated with transitions of molecules to discrete excited states and involve time intervals comparable with the period of the exciting radiation (10^{-14} sec) [Vavilov, (1950); Stepanov, (1955)].

Luminescence may be characterized by its integrated intensity and spectral composition. The dependence of the intensity I on wavelength λ is called the luminescence spectrum.

The fluorescence spectrum is always shifted toward longer wavelengths relative to the absorption spectrum (Stokes' law). However, the long-wavelength part of the absorption spectrum overlaps the short-wavelength part of the fluorescence spectrum to some extent. This is particularly characteristic of complex molecules. Fluorescence at wavelengths shorter than the wavelength of the exciting radiation is called anti-Stokes fluorescence. The additional energy necessary for its emission comes from an excess of vibrational energy in the excited molecules.

The dependence of the integrated intensity of luminescence, or the intensity of luminescence at a given wavelength, on the wavelength of the exciting radiation is called the luminescence excitation spectrum. The ratio of the number of emitted photons to the number of absorbed photons is defined as the quantum yield q. It has been shown experimentally for a number of substances that the quantum yield of fluorescence remains constant over a broad range of wavelengths; this is frequently referred to as Vavilov's law.

In general, an electronic transition will also involve a change in the vibrational and rotational energies of the molecule. The electronic spectrum is therefore in reality an electronic-vibrational-rotational spectrum. Since the energy levels have finite widths and tend to overlap in complex molecules, forming a continuous sequence, the profile of the electronic spectra of complex molecules has practically no vibrational structure. The shape of the absorption and emission bands is determined by the transition probabilities between different vibrational levels of the ground and excited states, and by the function describing the distribution of the vibrational levels over the energies of the upper and lower electronic states for absorption and emission respectively.

Shpol'skii Effect

The absence of a clearly defined vibrational structure in the electronic-vibrational spectra of complex molecules is a serious disadvantage in practical applications of molecular spectroscopy. Many attempts have therefore been made to find the conditions under which the vibrational structure of the spectrum can be made apparent. It is possible to reduce the number of degrees of freedom of molecules by freezing them in a solution. However, the solutions of many organic substances form molecular crystals in the solid state, and these have a number of specific absorptive properties leading to a broadening of the spectral emission band.

Intermolecular interaction can be suppressed almost completely by studying the substance under investigation in the gaseous phase. In this phase, however, there are many vibrational and rotational states forming a practically continuous sequence. This again leads to a broadening of the structure of the luminescence spectrum. Moreover, many complex molecules cannot be investigated in the gaseous phase without producing dissociation.

It is possible, however, to combine the advantages associated with cooling and vaporization, and at the same time avoid the complications introduced into the spectrum by the particular properties of these phase states. This can be achieved by placing the medium under investigation in another medium, known as the matrix or host, which ensures that the molecules are isolated from each other. The system is then frozen with the molecules fixed in definite orientations. The medium under investigation is thus really in the form of an "orientated gas" in the host medium.

Shpol'skii and co-workers [Shpol'skii and Il'ina (1951); Shpol'skii *et al.* (1952)] were the first to show that the luminescence spectra of frozen solutions of polyacenes, polyenes, and polyphenyls in normal paraffins as host media exhibit a vibrational structure, due to electronic transitions, which resembles line structure. (They referred to it as "quasi-line structure.") This phenomenon is sometimes called the Shpol'skii effect. According to Shpol'skii (1959, 1960, 1962, 1963), paraffin hosts can be used to prepare the medium under investigation (activator) in a state of molecular dispersion because the host and the activator are geometrically similar.

To ensure that the emitting molecules form an "oriented gas" the host must hold these molecules quite rigidly and must not be deformed by them. It must not interact with the luminescing molecules and must form with them a homogeneous solid solution. Finally, the host must be transparent throughout the region in which the luminescing molecules are excited.

By using *n*-paraffins as the host media, it has been possible to obtain quasi-line spectra for a broad range of materials, including, for example, the heterocyclic compounds [Nurmukhamedov *et al.* (1960); Shigorin *et al.* (1962)], some dyes of biological origin, e.g., porphyrins [Litvin and Personov (1961); Lyalin and Kobyshev (1963)]. However, quasi-line spectra have not thus far been obtained for protein, nucleic acids, and their components, although the appearance of such a spectrum would give information on the spatial relationships between the host and the embedded molecule, which is of particular importance for many aspects of molecular biology [Vol'kenshtein (1965)].

It is important to note that, apart from normal carbohydrates, it is also possible to use other solvents as host media, e.g., dibenzylaminoethanol [Val'dman and Sheremet'ev (1963)], cyclohexane [Kanda and Shimada (1961)], and benzene [Kanda *et al.* (1963)].

Extensive applications of the Shpol'skii effect in qualitative and quantitative analysis, in the study of the structure of polyatomic organic molecules, in crystal chemistry, and in solid state physics, were reviewed by Shpol'skii (1965*a*, 1965*b*).

Effect of the Solvent on the Electronic Spectra of Molecules

The intensity, the quantum yield, and the luminescence spectra are affected by the properties of the ambient medium. In this section we shall briefly review the effect of the solvent on the position of the spectra, a subject of considerable interest in connection with the main subject matter of this book.

Neporent and Bakhshiev (1960) have suggested the subdivision of molecular interactions causing shifts of electronic spectra in solutions into (1) universal interactions due to the effect of the properties of the solvent on the solute molecule, and (2) specific interactions between the solute molecule and one or a few of the surrounding solvent molecules due to the donor–acceptor interaction, association formation of hydrogen bonds, and so on.

The theory of the effect of universal interactions on the position of the electronic spectra of molecules [Bakhshiev (1961, 1964)] can be used to relate the shift of the spectra with the macroscopic parameters of the solvent, e.g., refractive index and dielectric constant, and the properties of the solute molecule, e.g., its volume and dipole moments in the ground and excited states.

Theory predicts the existence of four electronic levels (Fig. 0.2) for a molecule in solution: two equilibrium (A and B) levels and two nonequilibrium (A' and B') ones differing from the corresponding levels (E_0 and E_1) in an isolated molecule by the energy of interaction of the molecule with the surrounding solvent. Calling this energy W_i, we have:

$$W_i = \vec{\mu}_i \vec{F}_i \qquad (0.3)$$

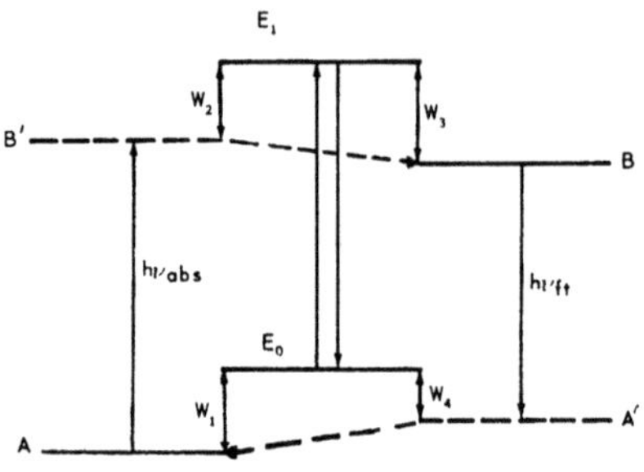

Fig. 0.2. Schematic illustrations of the electronic levels of a molecule in a solvent. Explanation in text.

In this expression $\vec{\mu}_i$ is the (permanent or induced) dipole moment of the molecule under consideration in the corresponding electronic state, and F_i is the electric field due to the whole set of surrounding solvent molecules in this state.

On excitation the electron density of the molecule undergoes a redistribution and its dipole moment is changed. The configuration of the solvent molecules has no time to change during the electronic transition, and the new state B' into which the molecule passes on absorption of a photon is a nonequilibrium one. Relaxation processes associated with a change in the configuration of the solvent molecules relative to the molecule in question then occur during the lifetime of the excited state. In the simplest case, when the relaxation time τ_r is much shorter than the lifetime of the excited state τ_{fl} (liquid solution), the molecule after emission passes into the equilibrium excited state B.

In the same way emission of a luminescence quantum brings the molecule into nonequilibrium ground state A'.

It should be noted that if $\tau_r \approx \tau_{fl}$ or $\tau_r \gg \tau_{fl}$ then the state B from which the emission takes place can also prove to be a nonequilibrium one. This case can occur in solid frozen solutions. The energy difference between the ground and excited states ΔW governs the relative shift of the absorption and fluorescence spectra. In general, this shift may be different from the analogous shift for fluorescence spectra in unfrozen solutions.

Polarized Fluorescence

The condition given by Eq. (0.2) is necessary but not sufficient to ensure that the molecule will absorb the incident light energy. It is also necessary that the electric vector in the light wave has a component parallel to the absorbing oscillator.

In practice, one has to deal not with an individual molecule but with an enormous number of such molecules. Since the molecules are not preferentially oriented in solution, absorption of light will be independent of its polarization. However, polarized luminescence can be observed even in the case of excitation by natural (unpolarized) light (spontaneous polarization). Luminescence with maximum polarization is then observed at right angles to the direction of propagation of the exciting radiation, because molecules whose oscillators are parallel to the direction of propagation of the incident beam (A_2 in Fig. 0.3) do not absorb the incident radiation. However, when luminescence is observed in the direction of propagation of the incident radiation, the polarization is not evident.

In addition to spontaneous polarization, one can investigate the polarization of luminescence excited by plane-polarized light. In this case polarized luminescence can be observed even in the direction of propagation of the incident radiation. Additional information about the position of the oscillators in space can be obtained by investigating the polarized luminescence of oriented molecules, e.g., in stretched films excited by plane-polarized light.

The degree of polarization can be defined by

$$P = \frac{I_1 - I_2}{I_1 + I_2}, \qquad (0.4)$$

where I_1 and I_2 are the intensities of components with perpendicular polarizations. The degree of polarization of luminescence depends on the mutual orientation of the absorbing and emitting oscillators in the molecule. Since emission usually occurs from a single excited state, the degree of polarization is usually constant throughout the luminescence spectrum.

The dependence of the degree of polarization on the wavelength of the exciting radiation is called its polarization spectrum. If the excitation of the molecule involves a single electronic transition the degree of polarization is independent of the wavelength of the incident radiation. When absorption involves the participation of a number of oscillators, the analysis of polarization spectra and comparison of such spectra with absorption spectra will provide information about the relative orientation of the absorption and emission oscillators. An analysis of this kind will show whether or not several electronic

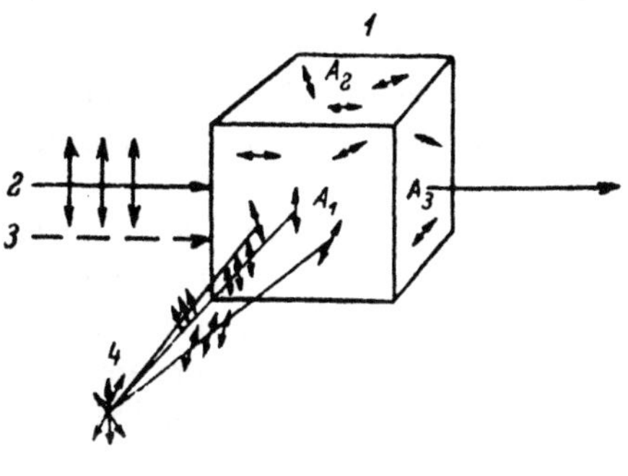

Fig. 0.3. Emission of polarized luminescence. (1) Container with solution; (2) linearly polarized exciting radiation; (3) natural light; and (4) partly polarized luminescence.

transitions take part in producing the absorption spectrum. Polarization spectra are frequently more specific for a given medium than are the fluorescence spectra.

The absolute value of the polarization is reduced by Brownian rotation of the molecules. The relationship between the degree of polarization, the properties of the luminescing molecules, and the properties of the ambient medium is given by the Perrin formula

$$\frac{1}{P}=\frac{1}{P_0}+\left(\frac{1}{P_0}-\frac{1}{3}\right)\tau\frac{RT}{V\eta}, \tag{0.5}$$

where R is the gas constant, T is the absolute temperature, η is the viscosity of the medium, V is the molecular volume of the luminescing molecules, τ is the lifetime of the excited state, and P_0 is the limiting polarization corresponding to $T/\eta \to 0$. Equation (0.5) can be used to determine the lifetimes of excited states if the molecular volumes of the luminescing molecules are known, and vice versa.

More detailed information about polarized luminescence can be found in the review and monograph by Feofilov (1948, 1961).

0.2. OPTICAL TRANSITIONS BETWEEN SINGLET AND TRIPLET STATES

It is well known that the electron has an intrinsic angular momentum (spin) which can assume one of the two values $\pm\frac{1}{2}(h/2\pi)$. According to the Pauli principle, a given orbit cannot contain more than two electrons with opposite spins. If all the spins in a molecule are antiparallel in pairs, the resultant spin S of the electrons is zero. If one of them changes its direction of spin, the resultant spin becomes equal to unity.

The quantity $2S + 1$ is the multiplicity of a given level, i.e., the number of possible orientations of the electron spin relative to the resultant angular momentum. When $S = 0$ we have $2S + 1 = 1$, and the corresponding energy level is called a singlet. When $S = 1$, so that $2S + 1 = 3$, the energy level is called a triplet. The ground (unexcited) state of a molecule is usually a singlet state. According to the Pauli principle, a change in the spin direction of one of the electrons is accompanied by a transition of the molecule to a triplet state. The appearance in the molecule of two unpaired electrons in the triplet state gives the molecule the properties of a chemically active biradical [Terenin (1947)].

It will be convenient to use the following designation of molecular energy levels: S_i for singlet and T_k for triplet; and $i = 0, 1, 2, 3,...$ and $k = 1, 2, 3,...$ to denote the position number of a level of given multiplicity (the state T_1 can be regarded as the ground state of the molecule with altered spin). Transitions between electronic levels are governed by the so-called selection rules, and can be divided into allowed and forbidden transitions. Allowed transitions are those in which there is no change in the multiplicity ($S_o \leftrightarrow S_1$, $T_1 \leftrightarrow T_2$ in Fig. 0.1):

forbidden transitions are those in which there is a change in the multiplicity ($S_o \leftrightarrow T_1, S_1 \leftrightarrow T_2$).

Singlet–singlet transitions are readily observed both in absorption and in emission. Radiation emitted as a result of transitions of the molecules from excited to ground singlet states is called fluorescence. In solutions, radiative transitions usually occur between the lowest excited singlet or triplet levels, even when higher-lying electronic levels are excited. This is often referred to as the Kasha rule [Kasha (1950)].*

Triplet–triplet transitions have been found in studies of the absorption of light by many different molecules [see, for example, McClure (1951)]. However, triplet–triplet fluorescence has been detected only in triphenylene solutions [Dupay *et al.* (1964)]. The lifetime of a molecule in a level T_2 is comparable with the lifetime in the first singlet excited level S_1.

Singlet–triplet transitions in solutions of polyatomic organic molecules have been observed both in absorption [West (1956)] and in emission [Parker (1964)]. Emission accompanying $T_1 \to S_0$ transitions is called phosphorescence. Optical transitions between terms of different multiplicity are forbidden in dipole emission (intercombinations are forbidden). The lifetime of a molecule in the triplet state for the triplet–singlet transition, which are forbidden, is considerably greater than the lifetime for allowed transitions.

In spite of the fact that they are forbidden, $S_0 \to T_1$ transitions occur because of the spin–orbital interaction. The level T_1 is not a pure triplet, but is in fact a superposition of triplet and singlet states. Spin–orbital interaction increases with increasing atomic number of the elements in the chromophoric group. The probability of $T_1 \to S_0$ transitions increases at the same time: for some compounds the probability of a phosphorescent transition is approximately proportional to the square of the spin–orbital interaction. This effect is readily seen by introducing heavy substituents, e.g., the halogens, into the molecule [West (1956); McGlynn *et al.* (1964)]. When spin–orbital interaction is strengthened, e.g., by the paramagnetic O_2 or NO molecules, this also leads to an increase in the probability of singlet–triplet transitions [Terenin (1947); Karyakin (1961); Barenboim (1963*a*)].

In spite of the fact that the probability of $S_0 \leftrightarrow T_1$ transitions can be increased for a number of materials, this probability is still small in comparison with the probability of allowed transitions. However, intersystem crossing results in a high population of the triplet level T_1 through the level S_1 (Fig. 0.1).†

*There are, however, a number of exceptions to this rule; e.g., azulene and diphenylene [McGlynn *et al.* (1964)]. Emission from S_2 as well as the main emission from S_1 has been found by Terenin and Ermolaev (1952) in chlorophyll excited by ruby laser.

†The levels S_1 and T_1 also have different multiplicities, but for a number of quantum-mechanical reasons [McGlynn et al. (1964)] the rate constants for intersystem crossing; e.g., $S_1 \leftrightarrow T_1$, are much greater than the rate constant for the $S_0 \to T_1$ electronic transition.

It is convenient to study $S_0 \leftrightarrow T_1$ transitions by investigating the associated phosphorescence because of the high population of the long-lived level T_1 (there is a large number of molecules in this state), the fact that phosphorescence continues even after fluorescence has decayed completely, and the fact that the two spectra are quite different.

Emission from the singlet excited state which is reached by the molecule after occupation of the triplet state ($T_1 \leftrightarrow S_1$, $S_1 \leftrightarrow S_o$) is called delayed fluorescence. This emission has the same spectrum as fluorescence, but has a different time constant. Detailed information about delayed fluorescence can be found in the review by Parker (1964). The account of delayed fluorescence given below is based on Parker's review.

Depending on the mechanism of the $T_1 \ldots \rightarrow S_1$ transition preceding delayed fluorescence, the latter can be divided into two types. In the first type, the transition $T_1 \ldots \rightarrow S_1$ occurs as a result of thermal activation of the intercombinational process $T_1 \rightsquigarrow S_1$. Accordingly, the emitted intensity is given by

$$I = Ae^{-E/kT}, \tag{0.6}$$

where T is the absolute temperature, k is Boltzmann's constant, and E is the activation energy necessary to take the molecule from the triplet state to the excited singlet state.

Since the intensity of this emission depends on the population of the triplet state, it is proportional to the first power of the intensity of the exciting radiation (monomolecular process). Other things being equal, the decay constant of delayed fluorescence is equal to the phosphorescence decay constant. This type of emission is occasionally referred to as *a* phosphorescence [Pringsheim (1949)].

The second type of delayed fluorescence occurs as a result of the so-called triplet–triplet annihilation in which the collision of two molecules of the same type in the triplet state results in an exchange of energy such that one of the molecules ends in an excited singlet state and the other in the ground state. A theoretical treatment of triplet–triplet annihilation has been given by Sternlicht *et al.* (1963). Triplet–triplet annihilation may be accompanied by the formation of triplet dimers and excimers, i.e., complexes consisting of an excited molecule and a molecule in the ground state [Parker, (1964)].

Since triplet–triplet annihilation is a bimolecular process, the intensity of the delayed fluorescence in this case is proportional to the square of the intensity of the exciting radiation. This is why the intensity decreases with decreasing concentration and increasing viscosity.

In addition to the above types of delayed fluorescence this phrase is occasionally also used to designate emission produced in recombinational processes following ionization. Further details will be found in Chapter 3.

0.3. LUMINESCENCE OF HETEROATOMIC MOLECULES

Electrons in the outer shells of atoms, which are responsible for the optical properties of molecules, are divided into three types: σ electrons which are responsible for ordinary bonding, π electrons which together with σ electrons form double or triple bonds, and n electrons (in atoms of N, O, S, P, etc.), which can participate in donor–acceptor interaction.

Transition of a molecule from one electronic state into another is associated with a change in the configuration of the electron shells. This change may involve different electrons in the molecule. $\pi^* \leftarrow \pi$ transitions occur if the excitation of the molecule involves a redistribution in the π-electron cloud. If the molecule contains heteroatoms, an n-electron from a nonbonding orbital localized on one of the heteroatoms may undergo a transition to an antibonding π-orbital [$\pi^* \leftarrow n$ transitions Kasha, (1950); Rao, (1961)].

Excited electronic states corresponding to these transitions are called $\pi\pi^*$ and $n\pi^*$ states. The classification of absorption spectra in accordance with the bands belonging to different electronic transitions can be found for different classes of chemical compounds in the book by Rao (1961). Absorption bands corresponding to $\pi^* \leftarrow \pi$ and $\pi^* \leftarrow n$ transitions are frequently found to overlap in the spectra. Absorption bands due to allowed $\pi^* \leftarrow n$ transitions are characterized by much lower molar absorption coefficients (ϵ = 300–2000) than the allowed $\pi^* \leftarrow \pi$ transitions ($\epsilon = 10^4 - 10^5$)[Kasha (1961)].

Oscillators associated with singlet $\pi^* \leftarrow \pi$ transitions lie in the plane of the aromatic rings, while oscillators associated with the triplet $\pi^* \leftarrow \pi$ transitions are perpendicular to the plane of the rings. Oscillators associated with $\pi^* \leftarrow n$ singlet transitions are perpendicular to the plane of the rings, while triplet transitions lie in this plane [Kasha (1961)].

The necessary (but not sufficient) condition for the presence of a $\pi^* \leftarrow n$ band in the absorption spectrum is the presence of a free pair of electrons (n electrons) in the heteroatom joined by a π bond to another atom in the molecule. Hence, one of the criteria for a band to belong to a $\pi^* \leftarrow n$ transition in the heteroatomic molecule is the absence of this band in the absorption spectrum of the hydrocarbon analog of the heteroatomic compound under consideration [Kasha (1961)].

We shall now consider the luminescence of heteroatomic molecules for different positions of the lowest singlet $\pi\pi^*$ and $n\pi^*$ levels (Fig. 0.4).

Case 1. The lowest singlet $\pi\pi^*$ level lies below the lowest singlet $n\pi^*$ level. Here the luminescence of the heteroatomic molecule is in fact similar to the luminescence of the polyatomic molecule not containing the heteroatoms. Depending on whether the excitation of the molecule occurs in the long-wavelength $\pi^* \leftarrow \pi$ or $\pi^* \leftarrow n$ absorption bands, one deserves $S_{\pi\pi^*} \rightarrow S_o$ fluorescence or $T_{\pi\pi^*} \rightarrow S_o$ phosphorescence.

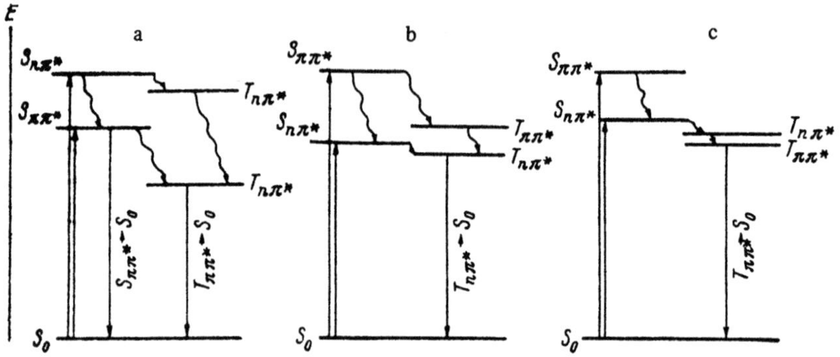

Fig. 0.4. Possible relative positions of the lowest $\pi\pi^*$ and $n\pi^*$ levels of heteroatomic molecules. S_0 indicate ground state; $S_{\pi\pi^*}, S_{n\pi^*}, T_{\pi\pi^*}, T_{n\pi^*}$ indicate lowest excited singlet and triplet $\pi\pi^*$ and $n\pi^*$ type, E is the state energy, (a) $E(S_{\pi\pi^*}) < E(S_{n\pi^*})$; (b) $E(S_{\pi\pi^*}) > E(S_{n\pi^*}), E(T_{\pi\pi^*}) > E(T_{n\pi^*})$; and (c) $E(S_{\pi\pi^*}) > E(S_{n\pi^*}), E(T_{\pi\pi^*}) < E(T_{n\pi^*})$.

Case 2. The lowest singlet $\pi\pi^*$ level lies above the lowest singlet $n\pi^*$ level. Here the molecule will not as a rule fluoresce [Kasha (1950, 1960)], and this is explained by the low probability of the optical transition $S_{\pi\pi^*} \to S_0$ and the high probability of conversion to triplet states. It is important to note that this rule may not be absolute. For example, Borresen (1963*a*) has found that pyrimidine exhibits fluorescence when the lowest excited singlet state is of the $n\pi^*$ type.

The luminescence of heteroatomic molecules is thus very dependent on the positions of the lowest $\pi\pi^*$ and $n\pi^*$ levels. In its turn, the energy of $\pi\pi^*$ and $n\pi^*$ levels is governed by the structure of the given molecule (in particular, the presence of substituents) and is very dependent on the state of the medium [Sidman (1958); Kasha (1961)].

In nonpolar solvents with neutral pH, the lowest excited singlet state of heteroatomic molecules is usually the $n\pi^*$ level [Kasha (1961)].

The attachment of different substituents to the molecule has an important effect on the energy of π^*–n transitions. The attachment of electron-donor substituents such as -OH, -SH, -NH_2 and so on, results in an increase in the energy of the lowest singlet $n\pi^*$ level [Kasha (1961)]. This is explained by the fact that the electron-donor substituents increase the density of the electron cloud in the system of conjugated bonds, e.g., in the ring of the heterocyclic molecule which impedes the transition of the n electron localized in the heteroatom to the π^* orbital. The influence of various substituents on the energy of $n\pi^*$ and $\pi\pi^*$ levels and the fluorescent properties of purines and pyrimidines has been discussed by Borresen (1963*a*).

An increase in the polarity of the medium will also give rise to a change in the energies of $\pi\pi^*$ and $n\pi^*$ levels. Thus, the $\pi^* \leftarrow n$ absorption band is shifted toward the shorter wavelengths as the polarity of the solvent increases (this is the so-called blue shift), while the $\pi^* \leftarrow \pi$ band is usually shifted toward longer wavelengths (red shift).

A change in the pH of the medium has a complicated effect on the relative position of the $\pi\pi^*$ and $n\pi^*$ levels. If the molecule contains one heteroatom which is responsible for the $\pi^* \leftarrow n$ absorption, protonation of this atom in an acid medium leads to the disappearance of the $\pi^* \leftarrow n$ absorption band [Kasha (1961)]. This is explained by the electrostatic interaction between the n electrons localized in the heteroatom and the proton which impedes the transition of the n electron to the delocalized π^* orbital. If several heteroatoms are present the molecule can have a number of $n\pi^*$ levels. Protonation of one of the heteroatoms may facilitate $\pi^* \leftarrow n$ transitions (reduce the energy of the $n\pi^*$ levels) due to other heteroatoms, because the attachment of a proton reduces the density of the electron cloud in the conjugated bond system and thus facilitates transition of the n electron to the delocalized π^* orbital.

The above mechanism for the effect of pH was put forward by Borreson (1963*a*) in order to explain the influence of pH on the fluorescence of purines and pyrimidines (see Section 2.1).

Thus, the relative energies of the $n\pi^*$ and $\pi\pi^*$ levels, and hence the luminescent properties of heteroatomic molecules, are exceptionally sensitive to changes in the conditions in the surrounding medium.

0.4. TRANSFER OF ELECTRONIC EXCITATION ENERGY. SENSITIZED LUMINESCENCE

A molecule can be excited either directly by the absorption of a quantum of energy or as a result of the radiationless transfer of an electronic excitation to another excited molecule:

$$D^* + A \rightarrow D + A^*$$

where D and A are the molecules donating and accepting energy, respectively. Another possible process is the absorption of energy by one chromophore in a molecule and emission by another group within the same molecule [Ermolaev and Terenin (1960); Lamola *et al.* (1965)].

Emission by the acceptor molecule whose excitation was produced by the radiationless transfer of energy from another molecule is called sensitized luminescence.

Depending on the type of interaction between the molecules there are three forms of radiationless transfer, namely, (1) induced resonance transfer, (2) exchange resonance transfer, and (3) excitation transfer. The relative importance of these mechanisms in the transfer of electronic excitation energy between molecules in solution is discussed, for example, in the review by Wilkinson (1964). In the induced resonance transfer of energy, the electromagnetic field associated with the electronic oscillator in the excited molecule (donor) induces forced oscillations in the unexcited oscillator in the acceptor molecule through a resonance interaction. This phenomenon may be compared with tuning-fork resonance except that the energy in the induced resonance interaction between molecules is transferred completely: only the donor or only the acceptor are excited at any given time. It is important to note that induced resonance transfer may occur at distances considerably greater than the linear dimensions of the molecules, i.e., there is no need for the electron shells of the donor and acceptor to overlap.

The theory of induced resonance transfer of energy was developed by Förster (1948, 1959), Galanin and Frank (1951), Dexter (1953), and Galanin (1960). The probability of transfer by the induced resonance mechanism is directly proportional to the overlap between the emission spectrum of the donor and the absorption spectrum of the acceptor, and decreases in proportion to the sixth power of the donor–acceptor separation. The distance R_0 between the donor and the acceptor at which radiative deactivation of the donor and the transfer of energy to the acceptor are equally probable is a measure of the likelihood of transfer by the induced resonance mechanism.

The only energy transfer processes which are allowed in the induced resonance interaction are those which are not accompanied by a change in the spin angular momentum of the donor and acceptor molecules [Wilkinson (1964)]. The singlet–singlet transfer D*(S) + A(S) → D(S) + A*(S) which leads to concentration quenching of fluorescence, concentration depolarization of fluorescence, the appearance of sensitized fluorescence, and the resonance quenching of fluorescence by impurities [Vavilov (1950)], satisfies this condition.

Induced resonance transfer may, however, occur even if it is apparently forbidden by spin selection rules when deactivation of the excited donor is also forbidden; thus, the radiationless triplet–singlet transfer D*(T) + A(S) → D(S) + A*(S) which was first predicted by Förster (1959) has been established experimentally by Ermolaev and Sveshnikova (1962, 1963).

The values of R_0 for the triplet–singlet transfer are close to the corresponding values for the singlet–singlet transfer. This is explained by the fact that in addition to the low probability of the triplet–singlet transfer, the probability of deactivation of the triplet donor molecule due to the emission of phosphorescence is also small: both processes are intercombinationally forbidden and therefore successfully compete with each other [Ermolaev and Sveshnikova (1963)].

In addition to the induced resonance interaction, an exchange resonance interaction may also lead to a radiationless transfer of energy. This occurs when the electron orbitals of the donor and acceptor molecules overlap [Dexter

(1953)]. In an exchange resonance interaction, the transfer probability decreases exponentially with distance, i.e., much more rapidly than in the case of induced resonance transfer.

Energy transfer processes in an exchange resonance interaction are allowed if the resultant spin angular momentum of the donor and acceptor is conserved (Wigner's rule). Thus, the radiationless transfer of an electronic excitation energy from a triplet donor molecule to an unexcited acceptor molecule, which is raised to a triplet level [Terenin and Ermolaev (1952); Ermolaev (1963)] in accordance with the scheme D*(T) + A(S) → D(S) + A*(T), is doubly forbidden in the induced resonance mechanism, but is completely allowed in the exchange resonance interaction. Triplet–triplet transfer is discussed by Ermolaev and Terenin (1960) and Ermolaev (1963).

0.5. LUMINESCENCE QUENCHING

The quantum yield of luminescence may be reduced (quenched) as a result of both intramolecular processes and intermolecular interaction between the luminescing molecule and the molecules in the medium. Let us consider the possible deactivation processes using the energy-level scheme shown in Fig. 0.1. In addition to radiative transition, a molecule may lose some of its electronic excitation energy through fast radiationless transitions between vibrational sublevels (vibrational deactivation), or it may undergo a transition from one excited state into another with the same multiplicity (internal conversion $S_1 \sim S_0$ or $T_1 \sim T_0$) having converted its electronic energy into vibrational energy. Finally, a molecule may undergo intersystem crossing, ending in an electronic state of different multiplicity and lower energy (Fig. 0.1). Internal conversion and intersystem crossing may be treated as an intramolecular energy transfer [Franck and Sponer (1956)].

Degradation of the excited state may also occur by expenditure of the available energy in a photochemical reaction in which the molecule dissociates by expending its excitation energy in breaking its own chemical bonds.

It is still not clear whether direct radiationless deactivation from the lowest excited singlet state is possible. Robinson and Frosch (1962, 1963) carried out a detailed theoretical analysis of energy degradation and concluded that degradation to the ground state is possible both from the triplet and singlet states. Ermolaev and Sveshnikova (1964) have shown experimentally that in most complicated aromatic molecules at low temperatures energy dissipation occurs largely through the triplet level.

Radiationless dissipation of energy in intercombinational conversion from an excited singlet to a triplet level ($S_1 \sim T_1$) explains the quenching of fluorescence with increasing temperature. Heating may lead to a reduction in the quantum yield in two ways: indirectly, through a change in viscosity of the solvent, and directly, through pure temperature quenching [Levshin (1951)]. In the second case it is assumed that the increase in the temperature gives rise to an increase in the probability of the intercombinational conversion $S_1 \rightsquigarrow T_1$. This process has an activation energy ΔE which is equal to the energy difference between the lowest level of the first singlet excited state and the point of intersection of the potential energy curves for the singlet S_1 and triplet T_1 states (Fig. 0.5). The quantum yield of fluorescence in the presence of temperature quenching is then given by

$$q = \frac{k_1}{k_1 + k_2 + k_3 e^{-\frac{\Delta E}{kT}}}, \qquad (0.7)$$

where k_1 is the probability of emission of fluorescence, k_2 is the probability of all radiationless processes in the absence of temperature quenching, $k_3 \exp(-\Delta E/kT)$ is the probability of temperature quenching, k is Boltzmann's constant, and T is the absolute temperature.

Quenching may be produced by different materials which either influence the rate constant for the intramolecular energy conversion or act as energy acceptors in an intermolecular energy transfer. Quenchers of the first type are called nonresonance quenchers, whereas quenchers of the second type are called resonance quenchers.

The mechanisms responsible for the effect of a nonresonance quencher on the rate of intramolecular conversion are different in different cases, but a necessary condition is always that the excited molecules of the fluorescent medium collide with the quenching molecules [Vavilov (1936, 1950); Sveshnikov (1952)]. Molecular diffusion occurring during the lifetime of the excited state is therefore of importance for this type of quenching.

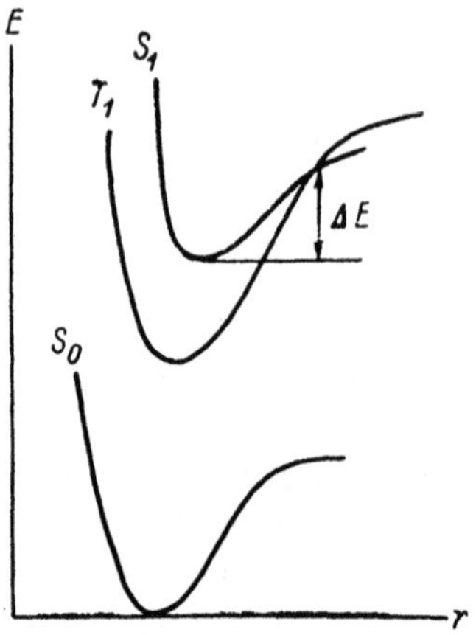

Fig. 0.5. Mechanism of temperature quenching of fluorescence S_0, S_1, and T_1 are the potential energy curves of singlet levels, ground and lowest excited and lowest triplet states, respectively. ΔE is the activation energy for the $S_1 \rightarrow T_1$ transition.

According to the theories of Vavilov and Sveshnikov, quenching of this kind is described by

$$\frac{q_0}{q} = (1 + 4\pi\rho DR\tau_0 C)\,\delta, \quad (0.8)$$

where q_0 and q are the quantum yields in the absence and presence of the quencher, respectively, τ_0 is the mean lifetime of the excited state of the fluorescing molecules in the absence of quenching, R is the sum of the radii of the fluorescing molecule and the quenching molecules, D is the diffusion coefficient, ρ is a coefficient which takes into account the probability of collisional quenching, and δ is a factor somewhat greater than unity and not very dependent on the concentration of the quencher.

For a resonance quencher the absorption spectrum overlaps the emission spectrum of the luminescing medium. The quenching action of such materials is due to the radiationless migration of energy between the luminescent molecule (donor) and the quenching molecule (acceptor) in accordance with the induced resonance mechanism (see Section 0.4). Sensitized luminescence by the quenching molecules may then arise. The theory of resonance quenching by impurities was developed by Förster (1948), Vavilov (1950), and Galanin (1955). Resonance quenching does not involve collisions between the luminescent molecules and the quenching molecules and is therefore independent of the viscosity of the solution.

A special case of resonance quenching is concentration quenching [Vavilov (1950)]. The quantum yield of luminescence begins to decrease for a given material beginning with a certain value of the concentration. Concentrational quenching may also be due to the formation of complexes consisting of excited and unexcited molecules.

Quenching due to transfer may also occur in other ways, namely, through the exciton and exchange resonance mechanisms.

Chapter 1

PHOTOLUMINESCENCE OF AROMATIC AMINO ACIDS, SYNTHETIC POLYPEPTIDES, PROTEINS, AND CELLS

The photoluminescence of proteins is due to the component amino acids (intrinsic luminescence) or various prosthetic groups if such groups are present in the protein molecule. To investigate structure, one can also examine luminescence protein by means of artificially attached groups (dyes) in which the dye luminesces (secondary luminescence).

In this chapter we shall consider only the intrinsic luminescence of proteins and the luminescence of aromatic amino acids responsible for it. We shall be largely concerned with factors governing the luminescent properties of these substances, and with the connection between protein structure and luminescence. We shall not be concerned with the classical applications of luminescence for analytical purposes. Readers interested in this aspect of the subject are referred to the monograph by Udenfriend (1962).

Proteins are linear copolymers of amino acids.

$$(H_2N{-}\underset{}{\overset{\displaystyle R}{\overset{|}{C}}}H{-}COOH),$$

joined by peptide bonds (–CO–NH–):

$$\ldots{-}\overset{\displaystyle R_i}{\overset{|}{C}}H{-}CO{-}NH{-}\overset{\displaystyle R_j}{\overset{|}{C}}H{-}CO{-}NH{-}\overset{\displaystyle R_k}{\overset{|}{C}}H{-}\ldots$$

There are about twenty most frequently encountered amino acids in proteins, differing from each other by the radical R. Structural formulas and pK values for these amino acids, which we shall need below, are given in Table 1.1. In contrast to the usual copolymers of nonbiological origin, proteins are descrete chemical substances with a definite complement and sequence of amino acid residues (primary structure of protein).

Table 1.1. Chemical Structure and pK Values of Some Amino Acids

Amino acid	Symbol	Chemical formula	pK_1	pK_2	pK_3
Tryptophan	Trp	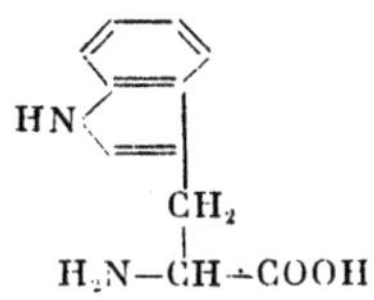	2.38	9.39	
Tyrosine	Tyr	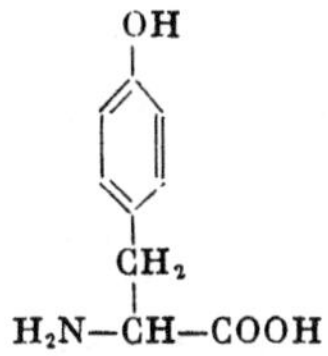	2.20	9.11	10.07 (OH)
Phenylalanine	Phe	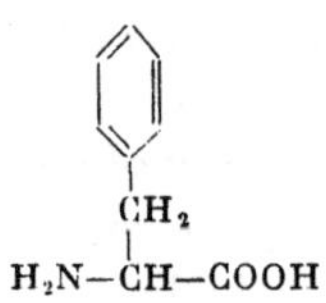	1.83	9.13	
Glycine	Gly	H H_2N–CH–COOH	2.34	9.60	
Leucine	Leu	H_3C CH_3 CH CH_2 H_2N–CH–COOH	2.36	9.60	
Glutamic acid	Glu	CH_2–COOH CH_2 H_2N–CH–COOH	2.10	9.47	4.07 (COOH)
Lysine	Lys	CH_2–NH_2 CH_2 CH_2 CH_2 H_2N–CH–COOH	2.18	8.95	10.53 (NH_2)

Note: pK_1 and pK_2 characterize the dissociations of COOH and NH_2 groups, respectively: the pK_3 refer to the dissociation of groups shown in parentheses. All these pK values are taken from Straub (1961).

Chemical bonds along a polypeptide chain are single bonds (apart from the peptide bond which is partly conjugated) allowing free rotation about them. This does not, however, lead to the reduction of the polypeptide chain into a statistical "knot" in the case of native protein. The formation of hydrogen bonds between the carboxyl and amide groups in the polypeptide chain stabilizes the relative disposition of its members, and this results in an ordered structure – the secondary structure of protein.

Different conformations are possible in the polypeptide chain. These include the α helix in which the amino acid residues are arranged in strict order along the helix so that each residue is joined by two hydrogen bonds to the fourth subsequent residue along the polypeptide chain. The α-helix model was originally put forward on theoretical grounds, and the structure was subsequently confirmed experimentally by X-ray studies of proteins and synthetic polypeptides.

The secondary structure is not necessarily a helical structure. Thus, the so-called cross β structure, in which hydrogen bonds join neighboring segments of a zigzag shaped polypeptide chain, has been found in synthetic polypeptides. In all the cases considered above the secondary structure is stabilized by intramolecular hydrogen bonds and may be stable in solution. An example of the secondary structure appearing as a result of the presence of intermolecular bonds is the β structure in which polypeptide chains arranged in a zigzag pattern and parallel to each other are joined by hydrogen bonds. It is important to note that the polypeptide chain can never be completely ordered. The secondary structure of the polypeptide chain is a combination of ordered and disordered segments.

Apart from the primary and secondary structure, proteins may also have a tertiary structure. This term is used to signify the packing of a helix or other structure in the protein chain into a globule. Stabilization of the tertiary structure (globule) is governed by various forces. These include the disulfide bonds between cysteine residues, hydrogen bonds between different groups in side chains, electrostatic forces between charged side and end groups, van der Waals forces due to the dipole moments or polarizability of side groups, and, finally, hydrophobic forces connected with the different "solubility" of side groups in the polypeptide chain. The last forces are responsible for the fact that polar groups interacting with water are found on the surface of the macromolecule, whereas nonpolar hydrocarbon radicals tend to collect inside the molecule.

The aggregation of several macromolecules as a result of the various interactions may lead to the appearance of a supramolecular quaternary structure. Detailed information about protein structure which will be useful in connection with the ensuing account can be found in the books by Bresler (1963) and Vol'kenshtein (1965).

1.1. LUMINESCENCE OF AMINO ACIDS

It will be shown later that the presence of the aromatic amino acid residues governs the luminescent properties of proteins and synthetic polypeptides.* Aromatic amino acids (tryptophan, tyrosine, phenylalanine) contain a system of delocalized π electrons in the aromatic nuclei and absorb light in the wavelength range 240–300 mμ (Fig. 1.1) when excited in this spectral region. They are found to fluoresce in solution [Teale and Weber (1957)] and in crystals [I. Ya. Barskii and E. M. Brumberg (1958)] at room and low temperatures [Vladimirov and Burshtein (1960)].†

The same amino acids emit phosphorescence in frozen solutions [Debye and Edwards (1952)] and powders [Nag-Chaudhuri and Augenstein (1964)] at liquid-nitrogen temperatures.

When amino acids are included in the protein molecule, their amino and carboxyl groups enter into a peptide bond. In this section, therefore, we shall pay particular attention to the effect of these groups on the luminescent properties of the chromophores of aromatic amino acids. Other factors, both intramolecular and those connected with the ambient medium, which also affect the luminescent properties of aromatic residues in the copolymer are discussed in sections concerned directly with protein luminescence.

Indole and Tryptophan

The emission of tryptophan is due to the presence of the indole ring

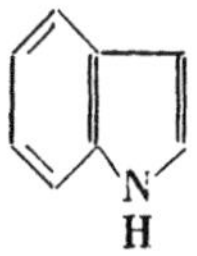

In neutral aqueous solutions, at room temperature, indole and tryptophan have similar continuous fluorescence spectra with a maximum around 350 mμ. The spectrum of tryptophan is shown in Fig. 1.2. The phosphorescence spectra of indole and tryptophan are best expressed in media having low polarity [Freed and Salmre (1958)]. The lifetime of the phosphorescence is 3 sec according to Debye and Edwards (1952), and 5.3 sec according to Longworth (1961). Moreover, Barenboim (1962) observed an after-emission from frozen solutions of tryptophan spectrally coinciding with its fluorescence and decaying exponentially with time (*a* phosphorescence).

*The luminescence of aliphatic amino acids [Rybac *et al.* (1955); Dimartin *et al.* (1957); Longin (1959); Douzou and Francq (1962)] and of the aromatic amino acids and proteins excited in spectral regions where there are no appreciable absorption bands [Steele and Szent-Györgyi (1957, 1958); Isenberg and Szent-Györgyi (1958); Fujimori (1960)] is interesting but still not well understood and will not be considered in this chapter.

†Tryptophan is a heterocyclic amino acid but in the literature on luminescence all the three amino acids – tryptophan, tyrosine and phenylalanine – are customarily referred to as "aromatic." We shall adhere to this usage here.

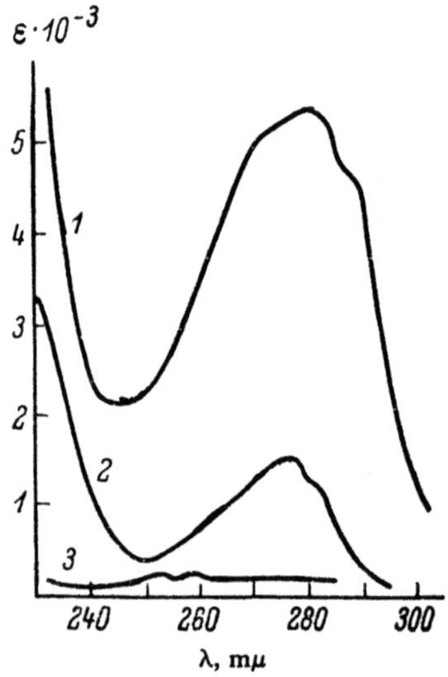

Fig. 1.1. Absorption spectra of neutral solutions of amino acids [McLaren and Shugar (1964)]. (1) Tryptophan; (2) tyrosine; (3) phenylalanine; ϵ is the molar absorption coefficient.

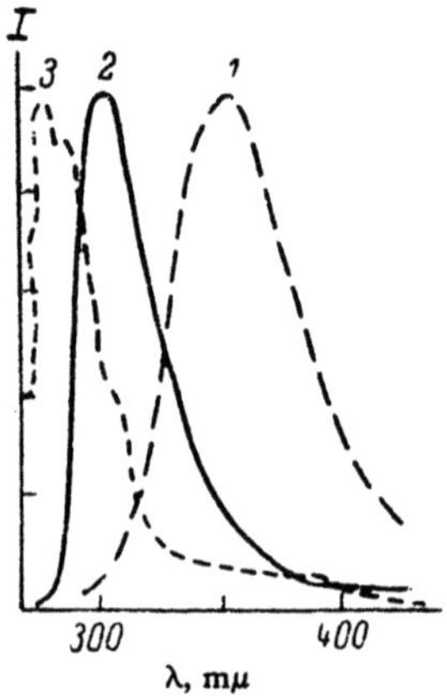

Fig. 1.2. Fluorescence spectra of some amino acids in neutral aqueous solutions at room temperature [Burshtein (1961)]. (1) Tryptophan; (2) tyrosine; (3) phenylalanine. I is the fluorescence intensity (rel. units.)

The fluorescence quantum yield of tryptophan (q = 0.2) in neutral aqueous solution is only half as large as that of indole (q = 0.4) owing to the influence exerted by the substituent group in the former compound on the fluorescence of the indole ring.

The influence of a wide range of substituents on the fluorescence quantum yield of the indole ring was studied by White (1959), by the authors of this book together with A. Sobota, and later, in greater detail, by Cowgill (1963*a*) (see Table 1.2). The latter author came to the conclusion that the effect is determined by the ability of the substituent to attract electrons, i.e., by the electro-

Table 1.2. Quantum Yields of Indole Derivatives in Aqueous Solutions

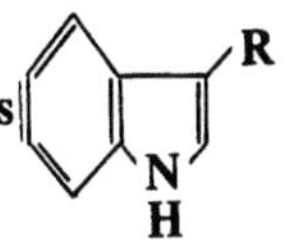

Compound	R	*q*
Indole	–H	0.40*; 0.40† 0.40‡
Indolyl-3-formic acid	$-COO^-$ $-COOH$	0.25* 0.08*
Indolyl-3-acetic acid	$-CH_2-COO^-$ $-CH_2-COOH$	0.45*; 0.47‡ 0.23*
Indolyl-3-propionic acid	$-CH_2-CH_2-COO^-$ $-CH_2-CH_2-COOH$	0.54*; 0.47‡ 0.37*
Indolyl-3-butyric acid	$-CH_2-CH_2-CH_2-COO^-$	0.48‡
Indolyl-3-lactic acid	$-CH_2-CH(OH)-COO^-$	0.41‡
Indolyl-3-acrylic acid	$-CH_2{=}CH-COO^-$	no fluorescence detected ‡
Glyceroindole	$-C(=O)-CH(OH)-CH_2(OH)$	no fluorescence detected ‡
Monoglyceride of indolyl-3-propionic acid	$-CH_2-CH_2-COO-CH_2-CH(OH)-CH_2-OH$	0.16‡
Monoglyceride of indolyl-3-butyric acid	$-CH_2-CH_2-CH_2-COO-CH_2-CH(OH)-CH_2-OH$	0.20‡
Tryptamine	$-CH_2-CH_2-NH_3^+$	0.40†
Tryptophan	$-CH_2-CH(NH_2)-COO^-$	0.51*
	$-CH_2-CH(NH_3^+)-COO^-$	0.20 §
	$-CH_2-CH(NH_3^+)-COOH$	0.09*

Compound	R	q
Tryptophanamide	$-CH_2-CH(NH_2)-CO-NH_2$	0.28*
	$-CH_2-CH(NH_3^+)-CO-NH_2$	0.14*
N-acetyltryptophan	$-CH_2-CH(NH-CO-CH_3)-COO^-$	0.28* 0.25† 0.28‡
	$-CH_2-CH(NH-CO-CH_3)-COOH$	0.128*
Ethyl ester of tryptophan	$-CH_2-CH(NH_2)-COO-C_2H_5$	0.076*
	$-CH_2-CH(NH_3^+)-COO-C_2H_5$	0.032*
Methyl ester of tryptophan	$-CH_2-CH(NH-CO-CH_3)-COO-CH_3$	0.11* 0.06†

*Cowgill [(1963*a*); corrected supplementary data to paper published in Biochim. Biophys. Acta, **133** (1): 6–18, 1967].

†White (1959).

‡ Work carried out by the present authors jointly with A. Sobota. It was assumed that the quantum yield $q = 0.20$ for tryptophan corresponds to a temperature of 20°C.

§ This yield was determined by the absolute method of Teale and Weber (1957). The remaining yields listed use the yield of tryptophan as a reference.

negativity of the substituent: a higher electronegativity results in decreased electron density in the indole ring and in decreased values of the yield.

The foregoing observations indicate that the effect of the atomic groups determining the electronegativity of the substituent on the quantum yield of the chromophore should decrease rapidly as the distance of this group from the chromophore is increased. In full agreement with this conclusion, the quantum yield of indole–CH_2–COO^- (q = 0.38) approaches that of indole itself (q = 0.40), while in indole–COO^- the COO^- group causes an appreciable lowering (q = 0.25) [Cowgill (1963*a*)].

Since the electronegativity of a substituent depends also on the state of ionization, it also follows from Cowgill's hypothesis that changes in the ionization states of the NH_2 and COOH groups of tryptophan should be reflected in the fluorescence yield. It has in fact been found [White (1959); Vladimirov and Li Chin-ho (1962); Cowgill (1963*a*)] that the fluorescence yield of tryptophan depends on the pH of the medium. Figure 1.3 shows that in the acid range (curve 1, section *ab*) the fluorescence is half-quenched at pH 2.4. The pH corresponding to half-quenched emission will from now on be denoted by pK_{fl}. In the case under consideration pK_{fl} coincides with the pK of the carboxyl group (it may be recalled that pK is numerically equal to the pH at which the concentration of ionized and nonionized forms of the molecule are the same). An analogous coincidence is observed for pK of the amino group and the pK_{fl} of tryptophan in the alkaline region (pH 9.4; Fig. 1.3, curve 1, section *cd*). On the other hand, the methyl ester of N-acetyltryptophan

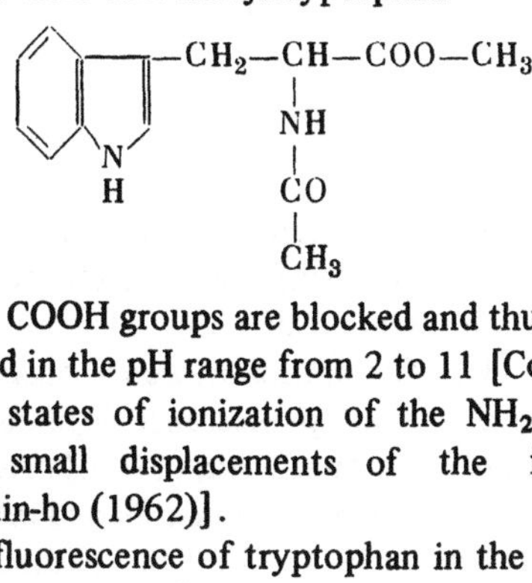

in which the NH_2 and COOH groups are blocked and thus cannot ionize shows a constant quantum yield in the pH range from 2 to 11 [Cowgill (1963*a*)].

Changes in the states of ionization of the NH_2 and COOH groups in tryptophan lead to small displacements of the fluorescence maximum [Vladimirov and Li Chin-ho (1962)].

The fall of the fluorescence of tryptophan in the range of pH corresponding to section *de* of curve 1 in Fig. 1.3 is ascribed to ionization of the imino

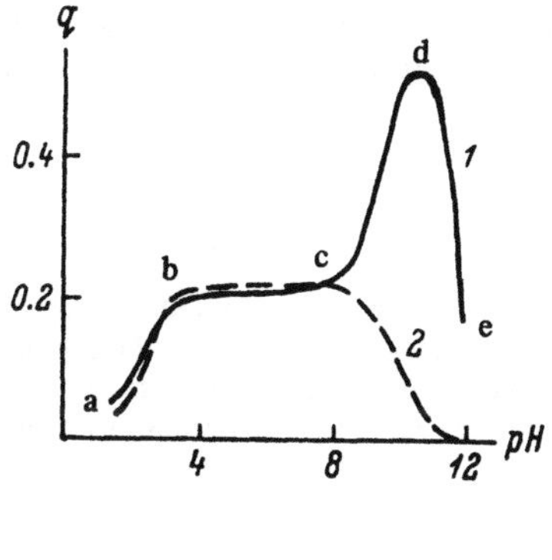

Fig. 1.3. Variation of the quantum yield of (1) tryptophan and (2) tyrosine with the pH of the medium [Cowgill (1963*a*)]. Explanation in text.

group of the indole ring. The effect of the NH group on the fluorescence of indole is confirmed by the fact that blocking of this group by methylation [Weber (1961*a*)] or reaction with formaldehyde [Konev and Chernitskii (1964)] eliminates the quenching at high pH.

Since the usual methods of titration did not reveal any ionization of the NH group of indole in alkaline media, White (1959) suggested that the ionized form appears only in the excited state ($R\text{–}NH \xrightarrow{h\nu} R^*\text{–}NH \xrightarrow{OH^-} R^*\text{–}N^- + H^+$) and that it is incapable of fluorescence. However, Konev and Chernitskii recently (1965) observed a weak fluorescence from indole (λ_{max} = 405 mμ) and tryptophan (λ_{max} = 420 mμ) in a strongly alkaline medium, and ascribed it to this ionic form: $R^*\text{–}N^- \rightarrow R\text{–}N^- + h\nu$. It may thus be concluded that ionization of the imino group does affect the luminescence of indole and tryptophan, the acid–base properties of which differ mainly in the excited states.*

Phenol and Tyrosine

The emission of tyrosine is due to the phenol ring

The fluorescence of phenol, tyrosine, and other phenol derivatives in neutral aqueous solution at room temperature is characterized by similar continuous spectra with maxima around 303 mμ . The fluorescence spectrum of tyrosine is shown in Fig. 1.2. The phosphorescence maximum of tyrosine lies at 387 mμ. The decay time of the phosphorescence depends strongly on the pH of the medium [Nag-Chaudhuri and Augenstein (1964)]. According to Debye and Edwards (1952) the decay time is 3 sec in neutral and 0.9 sec in alkaline media, while according to Longworth (1961) it is 2.1 sec. Although the quantum yields of phenol and tyrosine are very similar (respectively, 0.20 and 0.21), the examples show that q is considerably affected by the presence of substituents in the

---

It is sometimes stated [e.g., Vladimirov and Li Chin-ho (1962)] that pK_{fl} coincides with the pK of molecules in the excited state (pK). However, this is so far from being the rule, which may be illustrated by the following scheme:

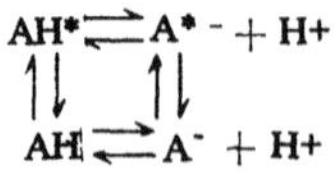

The pK_{fl} obviously depends on the relative rates of all the processes marked by arrows. When the rate of deactivation of the excited state is considerably faster than the rate of ionization in the excited state, the pK_{fl} should be the same as the pK of corresponding groups of the molecule in the ground state, In the general case pK_{fl} can assume any value between pK and pK*.

phenol ring (see Table 1.3). After measuring the quantum yields for a large number of phenol derivatives, Cowgill came to the conclusion that, as in the case of indole, the effect of the substituents depends on their electronegativity. The quantum yield of the fluorescence due to the phenol ring decreases with increasing electronegativity of the substituent, which in turn depends on the state of ionization. Ionization of the COOH and NH_2 groups of tyrosine should thus affect the fluorescence yield. The quantum yield of tyrosine at pH 4–8, where tyrosine is a zwitterion

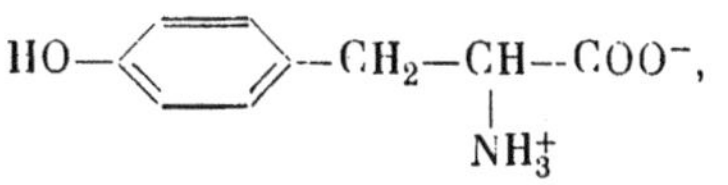

is equal to 0.21. The yield falls off together with pH, and at pH 2.5 becomes only about 0.1 (Fig. 1.3), in good correspondence with the pK of the dissociation of the carboxyl group.

Under alkaline conditions, the fluorescence quantum yields of phenol and its derivatives, in particular tyrosine, are influenced mainly by ionization of the OH group, leading to attenuation of the fluorescence. This is demonstrated by the good coincidence of the pK of the ionization of the phenolic hydroxyl and the pK_{fl} values of tyrosine and phenol in alkaline media. The fluorescence of anisole ($C_6H_5OCH_3$) does not show any changes in this region of pH [White (1959)]. Ionization of the OH group in tyrosine also results in spectral shifts. Cornog and Adams (1963) found a very weak fluorescence in tyrosine with an ionized OH group. The fluorescence maximum was considerably (by 30 mμ) displaced toward longer wavelengths in comparison with the maximum of the spectrum given by the un-ionized form.

On the basis of Cowgill's work, we should expect a certain increase in the q value of tyrosine following the dissociation of its NH_3^+ group, as was observed in the case of tryptophan. However, dissociation of the NH_3^+ group is not reflected by the pH dependence of the quantum yield of tyrosine [White (1959); Vladimirov and Li Chin-ho (1962)]. It may be supposed that the increase in the fluorescence quantum yield of tyrosine due to the dissociation of the NH_3^+ group is masked by the sharp fall of the yield due to ionization of the OH group.*

In one of his latest works, Cowgill (1965*b*) showed that dissociation of the NH_3^+ groups affects the pH dependence of the quantum yields of the dipeptides Gly–Tyr and Tyr–Gly. The variation of the fluorescence intensity with pH proved to be somewhat different in these two cases, which may be due both to differences in pK and to different distances of the amino groups in these compounds from the aromatic ring.

---

*The course of the quenching curve should depend on the relative concentrations of four forms of tyrosine distinguished by different combinations of OH, O^-, NH_3^+, NH_2 groups depending on the pH of the medium and the pK values of these forms, measured in particular by Yamazaki *et al.* (1965).

Table 1.3. Quantum Yields of Phenol Derivatives in Aqueous Solutions

Compound	Structural formula	q
Phenol	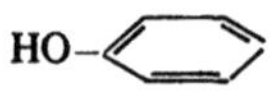	0.21* 0.21† 0.20‡
Anisole	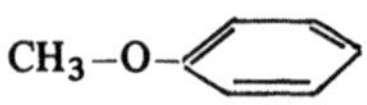	0.166‡
Tyramine	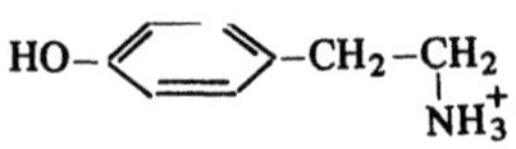	0.185† 0.21‡
Tyrosine	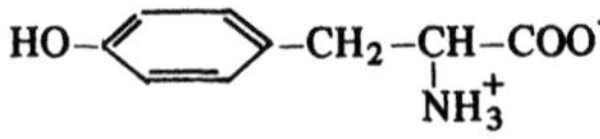	0.21† 0.21‡
	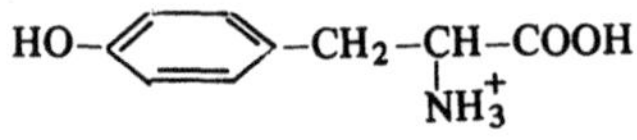	0.56†
N-acetyltyrosine	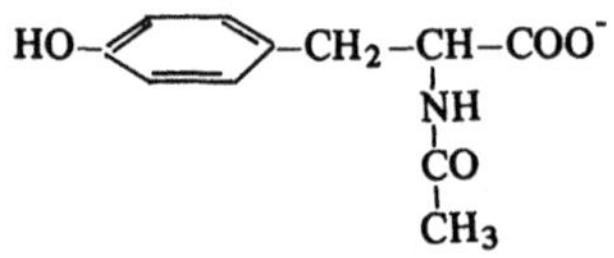	0.21‡
Methyl ester of tyrosine	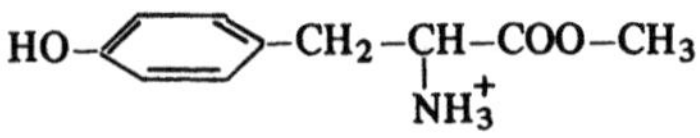	0.046‡
m-Cresol	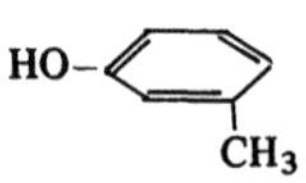	0.25†
m-Hydroxybenzyl alcohol	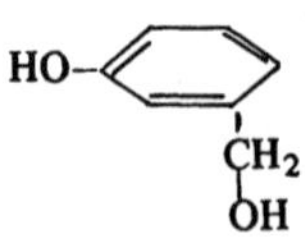	0.25†
m-Hydroxybenzoic acid	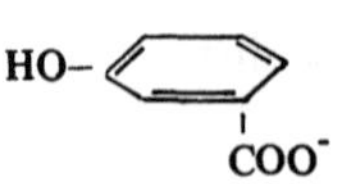	0.25†
	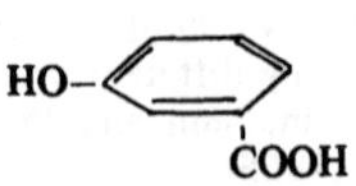	0.01†

Compound	Structural formula	q
m-Hydroxybenzaldehyde	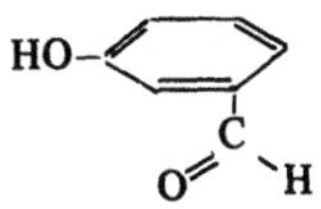	0.01†

*Yield determined by the absolute method of Teale and Weber (1957).
† Cowgill (1963a)
‡ White (1959)

Phenylalanine

The emission of phenylalanine is due to the benzene ring. In contrast to the other aromatic amino acids, phenylalanine has a structured spectrum (maxima at 282 and 289 mμ and "shoulders" at 298 and 305 mμ) which is close to being a mirror image of the absorption spectrum. The quantum yield of phenylalanine, which is only 0.04 in neutral aqueous solution, varies with the degree of ionization of the NH_2 and COOH groups [Vladimirov and Li Chin-ho (1962); Feitelson (1964)]. The fluorescence of the phenylalanine residue practically does not appear in proteins (see Section 1.2), and consequently we shall not be dealing with the luminescent properties of this amino acid, which are of little interest to the following discussion concerned chiefly with biopolymers. Data on the luminescence of phenylalanine may be found in several publications [Burshtein (1961); Udenfriend (1962); Feitelson (1964); Nag-Chaudhuri and Augenstein (1964); and Yeargers and Augenstein (1965)].

★ ★ ★

We have thus seen that the fluorescence of the aromatic amino acids tryptophan, tyrosine, and phenylalanine is due to the presence of indole, phenol, and benzene rings in these compounds. The quantum yields depend on the state of ionization of the OH group in tyrosine, the NH group in tryptophan, and the COOH and NH_2 groups in all these amino acids.

It has been proposed that the mechanism of the influence of the amino and carboxyl groups on the fluorescence of amino acids is determined by the electronegativity of these groups. Although this point of view, developed by Cowgill, finds a number of experimental confirmations, it is not always applicable and is not unanimously accepted by all authors. Thus, Feitelson (1964) considers that groups separated from the chromophore by several single bonds cannot exert any appreciable inductive effect on the fluorescent properties of the chromophore. According to Weber and Feitelson the reduction of the fluorescence quantum yields of tryptophan [Weber (1961a)], tyrosine, and phenylalanine [Feitelson (1964)] in acid media is due not to the inductive effect of this group but to any direct interactions with the aromatic ring leading to quenching (see later in this chapter). In addition, Cowgill himself (1963*a*) points out that if only the electronegativity of the substituent groups is taken into account it is difficult to see why compounds such as Leu–Tyr and Gly–Tyr

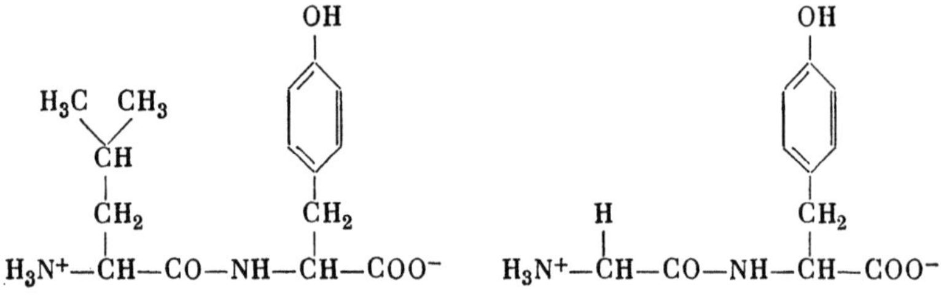

have different quantum yields (0.103 and 0.070 respectively).

Whatever the reasons for the effects exerted by the COOH and NH_2 groups on the fluorescence of the aromatic amino acids, it is clear that the fluorescence properties of proteins cannot be discussed simply on the basis of the corresponding properties of free amino acids in solution. In a protein the acids form part of the polypeptide chain and have no free amino and carboxyl groups, the very groups whose state of dissociation determines the quantum yields of these amino acids in solution.

On the other hand, the factors influencing the NH group in tryptophan or the OH group in tyrosine, or acting directly on the chromophore ring, may exert a similar effect on the fluorescence of amino acids in solution and in the polypeptide chain.

1.2. GENERAL CHARACTERISTICS OF THE LUMINESCENCE OF PROTEINS

Although the fluorescence of tissues and protein-rich organs was noticed over 100 years ago, and the luminescence of myosin was studied qualitatively in 1883 [Burshtein (1964)], quantitative work on the luminescence of proteins is relatively recent, dating back to the 1950's, when photomultipliers became available.

It has now been shown that nearly all proteins investigated luminesce at or below room temperature on excitation in the UV region [see the reviews by Burshtein (1961), (1964); Weber (1961*a*); and Udenfriend (1962)]. The only cases where luminescence has not been detected are the protamines (clupeine and sturine), whose molecules do not contain aromatic amino acids. These and other observations have led to the conclusion that the UV luminescence of proteins is due to tryptophan, tyrosine, and phenylalanine. The same three acids are responsible for the absorption spectra of proteins at 240–300 mμ. Detailed information about the absorption spectra of proteins may be found in a number of reviews [Beaven (1962); Wetlaufer (1962); McLaren and Shuger (1964)].

Fluorescence

The contribution made by the aromatic amino acids to the fluorescence of proteins is not always the same, and consequently proteins are commonly divided into two classes according to their luminescence properties: those containing tryptophan and those containing only tyrosine and phenylalanine as the aromatic amino acids.

Tryptophan-Containing Proteins. In these the fluorescence is determined mainly by the tryptophan residues, though other aromatic amino acids may also be present. This is true even for a protein such as human serum albumin, containing 17–18 tyrosine residues and 33 phenylalanine residues for every residue of tryptophan. The fluorescence spectra of native tryptophan-containing proteins are shifted towards the short wavelength region with respect to the spectrum of free tryptophan in solution, by amounts that are different in different cases. For example, in aqueous solution at room temperature the fluorescence maximum of trypsin lies at 332 mμ, that of chymotrypsin at 334 mμ, and that of bovine serum albumin at 342 mμ.

Considerable differences are also found among the quantum yields of proteins belonging to this class: at $\lambda_{exc} = 280$ mμ the quantum yield of gamma globulin is 0.038, while for bovine serum albumin $q = 0.152$ [Teale (1960)]. The quantum yield of the tryptophan-containing proteins increases appreciably when the excitation wavelength is changed from 280 to 300 mμ, as the proportion of light absorbed by tyrosine residues falls to zero [Teale and Weber (1959); Teale (1960)]. Taken together with the similarity between the excitation spectra of the proteins and tryptophan in solution, this is convincing evidence for the predominant part played by tryptophan residues in the fluorescence of proteins of this class. However, tyrosine too exerts some small influence on the fluorescence of the tryptophan-containing proteins. Weber (1961c) found two components in the fluorescence spectra of human and bovine serum albumins and in the spectrum of ovalbumin, using a method which he specially developed (1961*b*) for the analysis of fluorimetric data. The main component was due to tryptophan, and the second, with a maximum at 305±3 mμ and q = 0.018–0.022, to tyrosine. The presence of a tyrosine component in the fluorescence spectra of the tryptophan-containing proteins has also been confirmed by Bobrovich and Konev (1965).

Proteins Not Containing Tryptophan. The fluorescence of these proteins is determined by the tyrosine residues, as may be seen by the similarity between their fluorescence and excitation spectra and those of free tyrosine in solution. The quantum yields are low. For example, for the two protein trypsin inhibitors studied by Teale and Weber [Teale and Weber (1959); Weber (1961*a*)] q was found to be below 0.01. The largest fluorescence quantum yield reported for native proteins belonging to this class in aqueous solution, $q = 0.037$, is that of insulin [Teale (1960)]. The overall quantum yield of a protein is determined by the quantum yields of all the tyrosine residues: if a protein contains several such residues, their q values may differ appreciably from one another. Thus, it appears that in ribonuclease three tyrosine residues have $q = 0.055$, while for the remaining three $q < 0.002$ [Cowgill (1965*b*)]. The value of 0.055 is the highest of those found for tyrosine residues in proteins at neutral pH. Such a yield is found for the low-molecular polypeptide oxytocin and ribonuclease denatured by treatment with urea and mercaptoethanol [Cowgill (1964*a*)].

The fluorescence quantum yields of the tyrosine residues in proteins are, therefore, low regardless of whether tryptophan residues are present as well.

No phenylalanine component has ever been detected in the fluorescence spectra of proteins, perhaps because of the relatively low absorption coefficient and low quantum yield of this amino acid.

Phosphorescence

Protein solutions frozen down to temperatures of liquid nitrogen exhibit both fluorescence and phosphorescence. The latter phenomenon too is determined by the residues of aromatic amino acids [see the review by Burshtein (1961)].

The work on the phosphorescence of proteins received a strong impetus from the ideas of Szent-Györgyi (1957) about the important part played in various life processes by excited electronic states, especially triplet states. However, the interest of biologists died down when it was subsequently discovered that excited electronic states can hardly be very important in "dark" processes, and as a result, while the number of publications on protein fluorescence increased sharply during the last two or three years, very few papers on protein phosphorescence can be found in the literature [Longworth (1961); Augenstein and Nag-Chaudhuri (1964); Nag-Chaudhuri and Augenstein (1964)]. These papers appeared after 1960 and are not covered by Burshtein's review. They are also too limited to show whether phosphorescence will prove a useful tool for the study of the structure and function of proteins.

1.3. FLUORESCENCE AND THE STRUCTURE OF PROTEINS

Even the foregoing brief outline of the luminescence properties of proteins demonstrates that the fluorescence spectra, and above all the quantum yields, of aromatic amino acids change considerably when these acids are condensed into the polypeptide chain of a protein. For example, the quantum yields of the tryptophan-containing gamma globulin and bovine serum albumin are 0.038 and 0.152, respectively, as opposed to 0.20 for tryptophan itself. An even greater difference is seen between free tyrosine (0.21) and ribonuclease (0.017) whose emission is caused by tyrosine residues. The present section will be concerned with those factors which are responsible for these differences related to the macromolecular structure of the proteins.

Inclusion of the Aromatic Amino Acids into the Polypeptide Chain

In this examination of the way in which inclusion of the aromatic amino acids into the protein polypeptide chains changes the luminescence properties of the indole, phenol, and benzene rings we shall deal only with those effects on the chromophores that are transmitted through the chemical bonds between neighboring amino acid residues. We shall thus treat an idealized polypeptide chain extended along the peptide bonds.

For example, consider the chain fragment

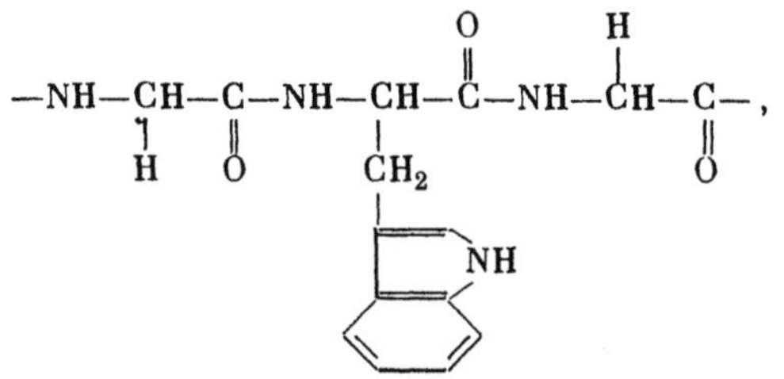

formed by tryptophan and glycine, the structurally simplest amino acid. In this chain the α-amino and the carboxyl groups (whose states of ionization determine the quantum yield of tryptophan in solution) enter into the peptide linkage, so that the fluorescence properties of the aromatic amino acids in solution and the same acids in a polypeptide chain will be very different. For this reason the use of the results obtained from studies of the fluorescence of mixtures of aromatic amino acids to analyze the luminescence of corresponding proteins seems to us unfounded, even if the relative concentrations of the amino acids are the same in the two cases, although this method has been used by Vladimirov and Burshtein (1960), Burshtein (1964), and certain other authors.

Table 1.4. Fluorescence Quantum Yields of Some Peptides [Cowgill 1963*a*)]

Substance	No. of peptide bonds	Quantum yield
Tyrosine	0	0.21
Gly–Tyr	1	0.07
Tyr–Gly	1	0.07
Gly–Tyr–glycinamide	3	0.035
Tryptophan	0	0.20
Gly–Trp	1	0.05

At first sight it may appear that the fluorescence properties of aromatic amino acid residues in a polypeptide chain should be similar to the properties of their chromophores (indole, phenol, and benzene) in solution, but this is not the case. Cowgill (1963*a*, 1963*b*) believes that the fluorescence of proteins is affected by the electronegativity of the peptide link, which in several cases may explain the low q values of the fluorescence of some proteins. This is demonstrated by the considerably lower quantum yields of various dipeptides and tripeptides in comparison with the free amino acids (Table 1.4.).

To analyze the influence of the peptide bond on the fluorescence of aromatic amino acid residues, it is interesting to look at the luminescence properties of glycyltyrosylglycinamide

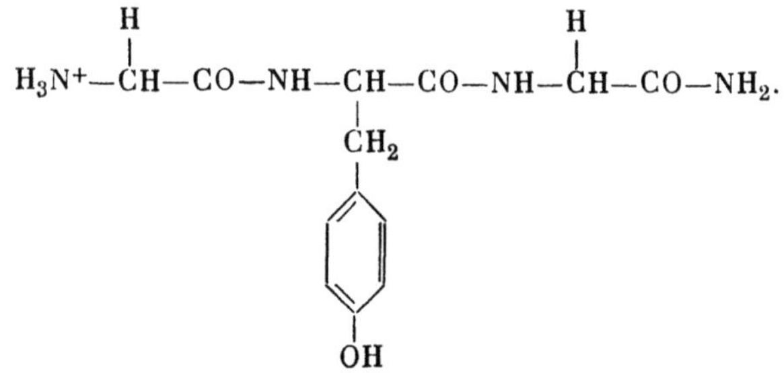

Electronegative groups capable of influencing the quantum yield of the tyrosine residue are here at a considerable distance from the ring, with the exception of the groups entering into peptide linkages. Therefore, this peptide may serve as a standard for the fluorescence of tyrosine included in a polypeptide chain. Its quantum yield is 0.035 [Cowgill (1963*a*)], which agrees in its order of magnitude with the quantum yields of tyrosine residues in proteins. However, in our opinion such a model has a serious shortcoming. The point is that in solution the molecule of glycyltyrosylglycinamide will have a certain configuration, and, in principle, the chromophore may find itself close to another part of the molecule, in particular to the quenching NH_2 group (see the following section).

The fluorescence properties of the aromatic amino acid residues in a protein and in free amino acids in solution are thus very different. A strong effect on the chromophores, leading to a lowering of their quantum yields, is exerted by the peptide bonds, so that the influence of the primary structure on fluorescence of chromophores should be much the same in various proteins. The very inclusion of aromatic amino acids into a peptide chain may explain the low quantum yields of some proteins.

The Influence of Quenching Groups

The side chains of a number of amino acids contain groups which are capable of quenching the fluorescence of nearby aromatic amino acid residues.

The prime examples of such groups are the carboxyls of aspartic and glutamic acids and the amino groups of lysine and arginine residues. This conclusion follows from experiments on the quenching of the fluorescence of aromatic amino acids in solution by substances containing amino or carboxyl groups [Teale (1960); Weber (1961*a*); Weber and Wold (1963); Feitelson (1964); Weber and Rosenheck (1964)]. The effectiveness and the mechanism of the quenching action of these groups depend on their state of ionization.

The Quenching Action of COO^- and NH_2 Groups. Substances containing these groups exert a strong suppressive action on the fluorescence yields of tyrosine and related components containing the OH radical [Feitelson (1964); Weber and Rosenheck (1964)]. On the other hand, the same groups have no effect on the fluorescence of phenylalanine and anisole (an O-methylated phenol). This indicates that the effect is associated with the presence of the OH group. Feitelson (1964) proposed that the presence of COO^- or NH_2 accelerates the rate of the dissociation of the OH group in the excited state, either by direct reaction with the hydroxyl proton of tyrosine or by catalyzing the dissociation.

The Quenching Action of COOH and NH_3^+ Groups. The quenching effect of acetic acid and hydroxylamine, which contain the above groups, on the fluorescence of indole and its derivatives was noted by Weber (1961*a*). Hydroxylamine quenches the fluorescence of indole and its derivatives much more effectively than the fluorescence of phenol derivatives.

The quenching of the fluorescence of the phenol derivatives by COOH was studied in detail by Feitelson (1964) and by Weber and Rosenheck (1964). Since carboxylic acids quench the fluorescence of phenol, anisole, tyrosine, and phenylalanine, the mechanism of their action does not depend on the presence of the OH group. Since the effect of the carboxylic acids decreased with rising temperature and is accompanied by increased solubility of anisole in water, and in view of the fact that at high COOH concentrations the yield tends to a constant value, Weber and Rosenheck (1964) proposed that the quenching is due to the formation of a "dark" complex between the aromatic nucleus of the fluorescent molecule and the acid.

Quenching of the Fluorescence of Aromatic Amino Acid Residues in Proteins and Synthetic Polypeptides

Since the action of the quenching groups considered occurs via OH or aromatic nuclei, the results obtained for free amino acids can obviously be extended to the amino acid residues in proteins.

The fluorescence of tryptophan residues in a protein may be quenched by COOH but not by NH_3^+, both potential proton donors. The fluorescence of tyrosine residues can be quenched either by proton donors (COOH and NH_3^+) or proton acceptors (COO^- and NH_2), though the latter are more effective; this follows from the experiments carried out on synthetic copolymers of glutamic acid and tyrosine [Rosenheck and Weber (1961); Fasman *et al.* (1964); Wada and Ueno (1964); Weber and Rosenheck (1964)] and of lysine and tyrosine [Pesce *et al.* (1964); Weber and Rosenheck (1964)].

The pH dependence of the quantum yield of –Glu–Tyr– (Fig. 1.4) may be explained as follows. In a neutral medium the fluorescence of the tyrosine residues is appreciably quenched by the side COO^- groups of glutamic acid; on passing to acid media, as the carboxyl groups become protonated this fluorescence is increased [Pesce *et al.* (1964); Weber and Rosenheck (1964)]. However, the cause of the fluorimetric titration is due not only to protonation of the carboxyl groups but also to structural changes of the copolymer caused by this protonation. In a neutral medium electrostatic repulsion between the negatively charged COO^- groups of glutamic acid residues prevents coiling into a helix, and the copolymer forms a random coil. Protonation of the COO^- leads to helix formation of the –L-Glu–L-Tyr– copolymer and to partial helix formation (< 20%) of the –L-Glu–DL-Tyr– copolymer [Doty and Gratzer (1962)]. It was found [Fasman *et al.* (1964)] that in a strongly acid medium, when all the carboxyl groups are protonated more fully, the helical form of–L-Glu–L-Tyr– has a fluorescence quantum yield around 12 times as large as that of the random coil, -L-Glu–DL-Tyr–. Several explanations may be proposed for this difference: (1) increased rigidity of the helical form; (2) differences in the polarity of the medium surrounding the tyrosine residues in the two forms [Fasman *et al.* (1964); see later in this chapter] and (3) differences in the accessibility of the tyrosine residues to the quenching COOH groups.

The fall in the fluorescence quantum yield of –Lys-Tyr– in the alkaline region (Fig. 1.4) is ascribed by Weber and Rosenheck (1964) to quenching of the tyrosine fluorescence by NH_2 groups due to dissociation of the NH_3^+ groups of the lysine residues.

Sensitivity of the fluorescence of aromatic residues to the state of ionization of the side groups in adjacent amino acids was observed in the study of a more complex model system, a copolymer of tyrosine, lysine, and glutamic acid [Rosenheck and Weber (1961); Weber and Rosenheck (1964)]. The pH dependence of the fluorescence yield of this copolymer is also illustrated in Fig. 1.4. The magnitude of q rises sharply at the isoelectric point.

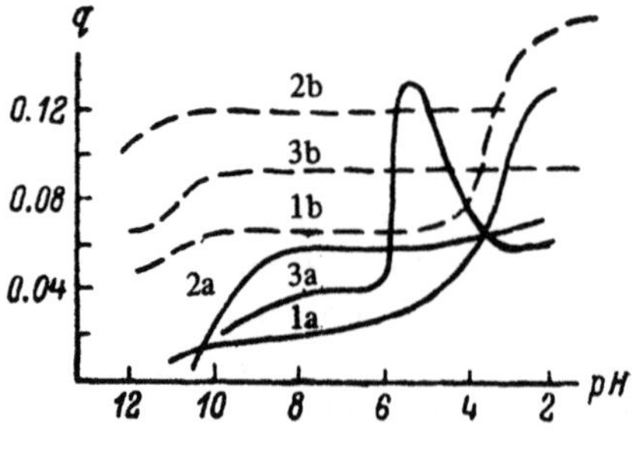

Fig. 1.4. Variation of the quantum yield q of copolymers with the pH of the medium [Weber and Rosenheck (1964)]. (1) Glutamic acid to tyrosine (96:4); (2) lysine to tyrosine (96:4); (3) glutamic acid to lysine to tyrosine (58:38:4); (a) before and (b) after O-methylation of tyrosine residues.

This form of dependence may be explained by the fact that at the isoelectric point the COO^- groups of glutamic acid combine with an equal number of the NH_3^+ groups of lysine, and consequently cannot exert their quenching action by interacting with the tyrosine hydroxyls [Weber and Rosenheck (1964)]. This hypothesis is confirmed by the absence of an isoelectric fluorescence maximum in salt solution and by the fairly high and constant value of q for the copolymer with O-methylated tyrosine residues.

The contents of the quenching amino and carboxyl groups are relatively much lower in proteins than in the copolymers discussed above, and therefore their effects cannot be expected to be equally pronounced. The probability of the mutual approach of aromatic rings and quenching groups should increase with the appearance of secondary and especially tertiary structure: packing of the polypeptide chain into a random globular coil leads to a higher content of quenching groups per unit volume.

Since the quenching action of amino and carboxyl groups depends on their state of ionization, it is natural to expect that their effects will appear in studies of the pH dependence of the quantum yields of proteins. However in an alkaline medium no effect of the dissociation of side NH_3^+ groups is observed, perhaps because the protein fluorescence is quenched by ionization of the tyrosine hydroxyls in more or less the same range of pH (a more detailed discussion of the quenching of protein fluorescence will be given in the section on migration of energy). In the case of tryptophan-containing proteins, whose structure is not altered in an acid medium, protonation of the carboxyls of glutamic and aspartic acids leads to a small fall of the fluorescence intensity below pH 5 [Steiner and Edelhoch (1963*a*)].

The side groups of some amino acids can thus quench the fluorescence of aromatic amino acid residues in proteins. It is quite possible that this effect is not confined to amino and carboxyl groups. Since the quenching requires a mutual approach of the quenching and fluorescing groups, the magnitude of the effect will depend on the actual structure of the protein.

The Effect of Hydrogen Bonding

The idea that the hydrogen bonds between tyrosine hydroxyls and charged carboxyl groups

$$-C_6H_4-OH \ldots {}^-O\!\!\!\!\!\!\quad\quad \genfrac{}{}{0pt}{}{{}^-O}{O}\!\!>\!C-$$

quench the fluorescence of tyrosine residues was put forward by Teale (1960, 1961) and Vladimirov (1961). The existence of such bonds is indicated most of all by the anomalous course of photometric titrations of proteins [see the review by McLaren and Shugar (1964)]. Teale's proposition was based on the fact that ribonuclease, in which three tyrosine residues out of the six are anomalous (hydrogen-bonded), has a lower quantum yield (0.017) than insulin (0.037) in which all the tyrosine residues behave normally.

Cowgill (1964*b*) established that in nonpolar media dimethylacetamide produces a strong quenching of the fluorescence of phenol also accompanied by a change in the absorption spectrum and ascribed this phenomenon to hydrogen bonding

$$C_6H_5{-}OH \ldots O{=}C\begin{matrix}\diagup CH_3 \\ \diagdown N(CH_3)_2.\end{matrix}$$

The quenching became weaker when the polarity of the medium was increased and disappeared completely in aqueous solution. On the basis of these results, Cowgill suggested that tyrosine residues situated in hydrophobic regions within a protein molecule may form hydrogen bonds not only with carboxyls but also with carbonyl groups, e.g., with the carbonyl groups of peptide bonds.

As could be expected, denaturation of ribonuclease leading to breakage of the hydrogen bonds between carboxyl (or carbonyl) groups and tyrosine residues increases the quantum yield to 0.055 [Cowgill (1964*a*)]. It has been shown [Cowgill (1965*a*)] that the three normal tyrosine residues of native ribonuclease have a yield of 0.055, the three remaining residues contributing not more than 10% to the total emission.

In our opinion there is at present no justification for a broader discussion of the influence of hydrogen bonding on the fluorescence of proteins.

The Effect of the Polarity of the Surroundings

The fluorescence of proteins is as a rule studied in polar solvents, most often water. However, the polarity of the medium immediately adjacent to the chromophores of the aromatic amino acids in the protein may differ appreciably from the polarity of the solvent, because these amino acid residues are screened from the solvent by the rest of the macromolecule. Such screening is very important in the presence of a globular tertiary structure, and can affect strongly the fluorescence properties of tryptophan residues. This is confirmed by the experimentally observed strong influence of solvent polarity on the fluorescence of indole [Van Duuren (1961); Weber (1961*a*); Mataga *et al.* (1964)] and its derivatives [Teale (1960); Van Duuren (1961); Steiner *et al.* (1964)]. The results obtained by Mataga and his associates (1964) are important for an understanding of the electronic spectra of indole (Fig. 1.5), tryptophan, and tryptophan residues in proteins.

In nonpolar media the fluorescence spectrum of indole is strongly structured and has the main maximum around 300 mμ. In polar media this maximum disappears and is replaced by a longer-wave unstructured fluorescence band, which moves progressively toward longer wavelengths with increasing polarity of the medium. The explanation given by the authors for this behavior is best illustrated by the scheme presented in Fig. 1.6. The absorption spectrum

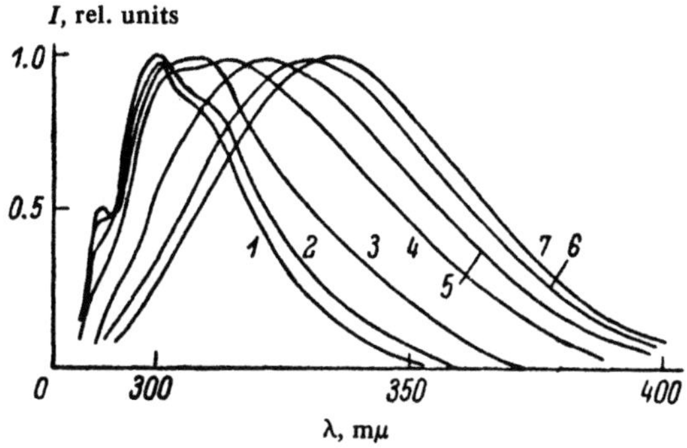

Fig. 1.5. Fluorescence spectra of indole in mixtures of *n*-hexane and ethanol [Mataga *et al.* (1964)]. Ethanol concentration (*M*): (1) 0; (2) 0.017; (3) 0.085; (4) 0.17; (5) 0.51; (6) 2.55; (7) 8.52.

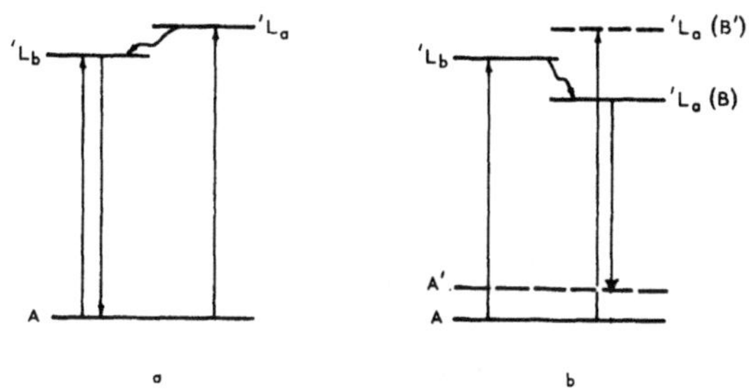

Fig. 1.6. The electronic levels of indole in (a) nonpolar and (b) polar solvents. The notation *A*, *A'*, *B*, and *B'* corresponds to that in Fig. 0.2.

of indole at 250–300 mμ consists of two electronic bands, $A \to {}^1L_b$ and $A \to {}^1L_a$, with maxima lying at 286 and 270 mμ respectively.

The fluorescence spectrum of indole in a weakly polar solvent is determined by an $A \leftarrow {}^1L_b$ transition.

The large dipole moment of indole in the excited ${}^1L_a$ state leads to a considerable lowering of the energy of this level as the polarity of the medium is increased (see p. 6). Inversion of the ${}^1L_a$ and ${}^1L_b$ levels occurs, and in a polar solvent the ${}^1L_a$ level becomes lower and responsible for the fluorescence.

Let us now turn to the influence of the screening of tryptophan residues from the solvent on the fluorescence of a protein. Different tryptophan residues will find themselves surrounded by media of different polarity, and will consequently have different spectral maxima, as can be seen by comparison with the above data for indole. The spectrum of each protein will therefore be the result of some averaging of these maxima. It is possible that the slight structure in the fluorescence spectra of tryptophan-containing proteins observed by Vladimirov and Burshtein (1960) is due to such superimposition of several spectra due to different tryptophan residues.

The individuality of the structure determining the polarity of tryptophan residues should result in differences between the values of the spectral maxima of proteins, and this has in fact been found in practice, for example by Teale (1960). Since screening reduces the polarity, the spectra of proteins will be shifted towards shorter wavelengths in relation to the spectra of indole and tryptophan [Edelhoch and Steiner (1964)].*

If proteins are treated with 8*M* urea (Teale, 1960) their fluorescence maxima become close to one another and to the maximum given by tryptophan in aqueous solution. This may be because 8*M* urea partly destroys the protein structure, making the tryptophan residues more accessible to the solvent (water). This in turn increases the polarity of the surroundings and results in a long-wave shift of the spectrum [Edelhoch and Steiner (1964)].

In this context, it is interesting to mention the work of Bobrovich and Konev (1964) who were able to obtain sharply structured fluorescence spectra of crystalline amylase (Fig. 1.7). One of the possible explanations proposed by these authors for the phenomenon is that it is similar to Shpol'skii's effect (Section 0.2).

The authors suggest several possible reasons for the appearance of such a spectrum, e.g., differences in the physicochemical character of the medium surrounding the tryptophan residues in crystalline amylase on the one hand and in amylase in solution or in the amorphous state on the other hand; and changes in intramolecular hydration, close contact and interpenetration of

*According to Vladimirov and Li Chin-ho (1962), the short-wave shifts of the fluorescence of tryptophan-containing proteins and tryptophan crystals in comparison with the free acid in solution are both due to a restriction of the vibrational degrees of freedom. The actual reason for the shift is thought to be hydrogen bonding between the tryptophan NH and adjacent acid groups.

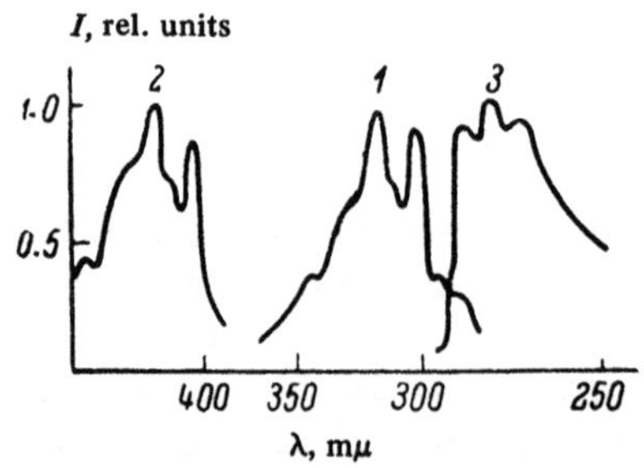

Fig. 1.7. Fluorescence (1), phosphorescence (2), and excitation (3) spectra of crystalline amylase [Bobrovich and Konev (1964)].

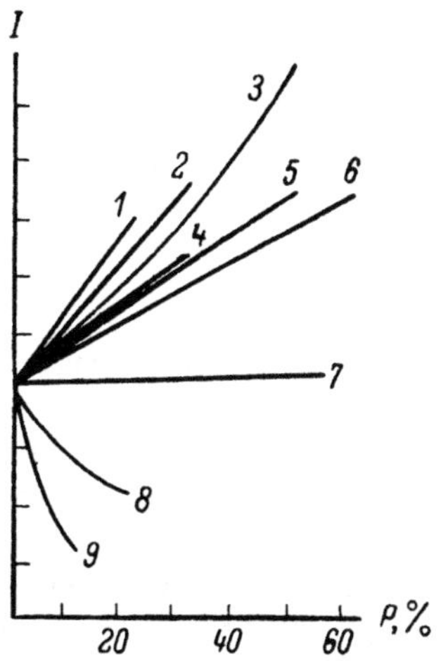

Fig. 1.8. Dependence of the fluorescence intensity (I) of acetyltryptophanamide in a phosphate buffer on the volume fraction (p) of various substances added to the solution [Steiner *et al.* (1964)]: (1) dimethyl sulfoxide; (2) sodium acetate; (3) dioxane; (4) potassium chloride; (5) propylene glycol; (6) urea; (7) sucrose; (8) glycine; (9) glycylglycine; (buffer pH 7.0; λ_{exc}, 290 mμ; λ_{fl}, 350 mμ).

polypeptide chains of neighboring molecules, and even reversible changes in the secondary and tertiary protein structure during crystallization. Practically nothing is known about the state of water in crystalline amylase. However, it may be supposed that some or all of the tryptophan residues are situated in crystalline amylase in a nonpolar medium. This point is supported by X-ray studies, which show that haemoglobin molecules in a crystal are associated with water only at the surface [cited by Eley (1962)]. Thus, on the basis of the above argument about the effect of the medium polarity on the fluorescence spectra of indole, we may postulate that the short-wave position of the fluorescence spectrum of crystalline amylase and its structure are largely determined by the transition $A \rightarrow {}^1L_b$. Bobrovich and Konev (1964) associate with this transition only the short-wave maxima at 288 and 293 mμ. The contribution of various oscillators to the fluorescence spectrum of crystalline amylase can probably be elucidated by studying polarization fluorescence spectra.

Apart from determining the position of the fluorescence spectrum, the polarity of the medium affects the quantum yields of indole derivatives. According to Teale (1960), a lowering of the dielectric constant of the medium leads, apart from a blue shift of the fluorescence maximum, to an increase in the quantum yield of glycyltryptophan. Steiner *et al.* (1964) noted that additives that decrease the dielectric constant of the solvent (propylene glycol, dioxane, sucrose, dimethyl sulfoxide) increase the fluorescence yield of acetyltryptophanamide. Conversely, glycine and glycylglycine, which raise the dielectric constant, lead to a quenching of the fluorescence. An exception to this qualitative rule is urea (Fig. 1.8).

The data on the effects of solvent polarity on the fluorescence of phenol and its derivatives are more contradictory. Weber and Rosenheck (1964) report that the polarity of the medium has practically no effect on the fluorescence spectra and yields of phenol and tyrosine. At the same time, Wada and Ueno (1964) find that the quantum yields of polytyrosine in ethanol, dioxane, and water are in the ratio of 15:2:1.

Structure of Proteins and Migration of Energy

The whole problem of the migration of energy in biological systems has been treated by a large number of authors, often with highly contradictory results. Detailed analyses of these results and bibliographies will be found in a number of reviews [Ril' (1948); Vladimirov and Konev (1957, 1959); Stryer (1960); Tarusov (1960); Burshtein, (1961, 1964)].

It has now been established that resonance migration of energy takes place in solutions, and distances R_o (Section 0.6) have been determined between molecules of phenol (17 Å), indole (17 Å), from phenol to indole (18.5 Å) [Weber (1960*a*)], and between tryptophan molecules (17 Å) [Konev *et al.* (1961*a*, 1964)]. Wada and Ueno (1964) demonstrated migration in model systems, namely, tyrosine and tryptophan copolymers. Resonance migration of energy in proteins has been shown to occur between tyrosine [Weber (1960*a*, 1960*b*, 1961*a*)] and tryptophan [Konev *et al.* (1965*a*)] residues, and between tyrosine and tryptophan [Bobrovich and Konev (1965)]. If the protein contains prosthetic groups the energy can migrate to these from aromatic amino acid residues [see, for example, Vladimirorov and Konev (1957, 1959)].

Finally, from the fluorimetric titrations of proteins carried out by Steiner and Edelhoch (1963*a*, 1963*b*) and by Cowgill (1965*b*), and of synthetic polypeptides [Cowgill (1963*b*, 1965*b*); Edelhoch *et al.* (1963)], it follows that ionized tyrosine residues are effective acceptors of the excitation energy of tryptophan and tyrosine residues in alkaline media.

It is not difficult to see that there should be a certain connection between the resonance migration of energy in a protein and its structure. In fact, since the probability of resonance migration decreases, under otherwise equal conditions, with the sixth power of the distance between the donor and the acceptor (see Section 0.4), the principal condition of energy migration between chromophores of the aromatic amino acid residues in a protein should be that they are physically close in space. In turn, the separation between the chromophores depends on the primary, secondary, and tertiary structure of the protein. Only specific features of the primary and secondary structure can ensure the necessary approach of certain aromatic residues. For example, the probability of energy transfer to an ionized tyrosine residue decreases progressively with increasing number of groups separating the tryptophan and tyrosine residues in compounds with different primary structure: for Trp–$(Gly)_n$–Tyr the migration probability is 0.85 with $n = 0$ and

0.35 with $n = 4$ [Edelhoch *et al.* (1963)]. The optimum condition for energy migration is a tertiary structure ensuring the closest packing of the amino acid residues.

The point now is, can one draw any conclusions about the structure and structural conversions in proteins on the basis of migration? An approach to the solution of this problem was made by Cowgill (1965*a*), who investigated the effect of iodination of tyrosine residues on the fluorescence of insulin, ribonuclease, and a number of indole compounds.

Bearing in mind that iodinated tyrosine residues ao not fluoresce and act as acceptors of the excitation energy of the noniodinated residues, Cowgill obtained the equation

$$I = I_0\left(1 - \frac{[p]}{m\,[M]}\right)\left(1 - k\frac{[p]}{[M]}\right), \tag{1.1}$$

in which I is the fluorescence intensity at various stages of iodination, I_0 is the intensity before iodination, and $[p]$ is the concentration of monoiodinated residues; because in the early stages of iodination the main product is the monoiodinated compound Tyr + I_2 = I–Tyr + HI. The symbol $[M]$ represents the concentration of protein molecules in the solution, m is the number of fluorescing tyrosine residues per molecule, and k is the effectiveness of the migration of energy from tyrosine residues to iodinated tyrosine residues. Thus $k = 1$ corresponds to 100% migration, i.e., to complete loss of fluorescence in the presence of one iodinated residue per molecule, and $k = 0$ indicates the absence of migration or the special case when the molecule contains only one tyrosine residue.

It was found that the experimental dependence of I on $[p]/[M]$ for the investigated substances is best described by the theoretical formula (1.1) for the following value of m and k:

	m	k
Tyr–Tyr	2	0.3
Insulin	4	0.3
Native ribonuclease*	3	0.2
Denatured (7.2*M* urea) ribonuclease	6	0.06

These results lead to a number of important conclusions [Cowgill (1965*a*)]: (1) the equal values of k for insulin and Tyr–Tyr show that in insulin four tyrosine residues are very close to one another; (2) in native ribonuclease only three out of six tyrosine residues are fluorescent; (3) denaturation of ribonuclease makes all six tyrosine residues "normal" and fluorescent; (4) comparison of the k values indicate that there is a large spatial separation of the tyrosine residues in ribonuclease after denaturation.

*It is assumed that the "normal" tyrosine residues of ribonuclease are iodinated first [Donovan (1963)].

Thus the effectiveness of energy migration can throw some light on the structural characteristics of proteins.

★ ★ ★

The material reviewed in this section indicates that the fluorescence properties of proteins depend on a whole series of factors which affect the fluorescence properties of the aromatic residues. The action of these factors is closely connected with the primary, secondary, and tertiary structure of the protein.*

It is fairly certain that the above-discussed factors are not the only ones to determine the fluorescence of aromatic amino acid residues in a protein. For example, our information is as yet too scant to enable us to estimate the possibility of the appearance of a fluorescent band with a maximum at 420–470 mμ which was found by Lehrer and Fasman (1964) for several helical synthetic polypeptides and which these authors ascribed to excimers of aromatic amino acid residues. Similarly, we do not believe that much can be said at present about the way in which fluorescence is affected by molecular rigidity due to the secondary and tertiary structure.

1.4. FLUORESCENCE INVESTIGATIONS AS A METHOD OF STUDYING THE STRUCTURE AND THE CONFORMATIONAL TRANSITIONS OF PROTEINS

Although the structure of a protein is determined by the mutual disposition of large sections of the macromolecule and its luminescence by the properties of small chromophoric centers, the fluorescence properties of a protein are closely connected with its structure. This was already shown in the preceding sections. We shall now attempt to assess the information about the structure and structual transformations of proteins that may today be gathered from a study of UV fluorescence.

Changes in Fluorescence Intensity as a Criterion of Conformational Transitions. Comparison with Other Methods

It has now been shown experimentally that the fluorescence intensity of proteins is highly sensitive to all structural changes. As an example of the application of fluorescent methods to the study of conformational transitions we can consider the experiments carried out by Perlman (1964) who studied in parallel the fluorescence intensity, optical rotation, and biological activity of pepsinogen (its tendency to transform into pepsin) as a function of

*Konev (1964) reports that changes in the quaternary structure are also reflected in changes of the fluorescence spectra, but these changes of the fluorescence intensity could be due to associated alterations in the secondary and tertiary structure.

temperature and concentration of urea. An increase of optical rotation and a reduction of fluorescence intensity occurs in the same temperature region (Fig. 1.9). In the same region the biological activity falls off with rising temperature. If pepsinogen is heated in the presence of urea the conformational transition takes place at a lower temperature. The transition temperatures (corresponding to the points where the plots in Fig. 1.9 bend over) determined from measurements of the optical rotation and fluorescence are in good agreement. Thus the changes in the above three characteristics reflect one and the same conformational transition in the macromolecule.

However, the correlation between the changes in optical rotation and in the fluorescence intensity is not complete; cooling of the pepsinogen solution after a single heating to 60°C is accompanied by a full recovery of biological activity and fluorescence, while the optical rotation is only partly restored. Repeated heating and cooling result in partial loss of the biological activity and fluorescence. According to Perlman, this relationship between the fluorescence intensity and biological activity indicates that some of the tryptophan residues responsible for the fluorescence are at the active center of the pepsinogen molecule.

Numerous publications appeared in recent years [Steiner and Edelhoch (1961, 1962*a*, 1963*a*, 1963*b*); Brand *et al.* (1963); Edelhoch and Steiner (1962); Gally and Edelman (1962, 1964); Gerstein *et al.* (1963); Cowgill (1964*a*); Steiner *et al.* (1964); Frattali *et al.* (1965)] in which the study of conformation transitions in proteins by the luminescence method is compared with the results obtained by other methods. Such comparisons show that fluorescence methods are both convenient and effective. However, as the structural changes of proteins are reflected indirectly on the fluorescence properties of the chromophores, it is at present impossible to obtain detailed information about the character of the transition from changes in the fluores-

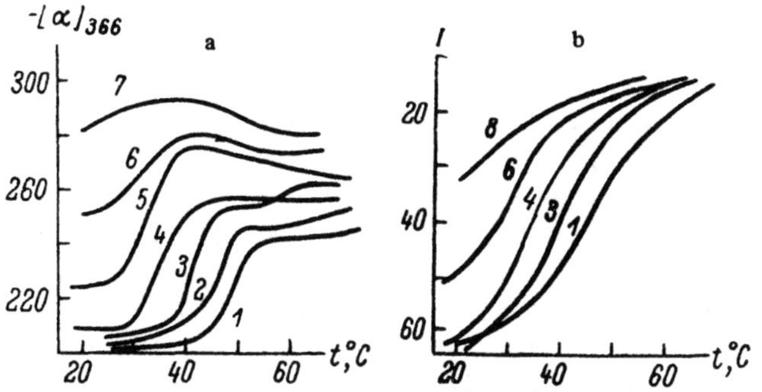

Fig. 1.9. Temperature variation of (a) the optical rotation and (b) the fluorescence intensity of pepsinogen solutions [Perlman (1964)]. Urea concentration, *M*: (1) 0.0; (2) 0.5; (3) 1.0; (4) 2.0; (5) 2.5; (6) 3.0; (7) 3.5; (8) 10.

cence intensity. Further development of the method will evidently make it possible to define more closely the local changes in the micro-surroundings of aromatic amino acid residues in proteins that occur as a result of conformational transitions.

Some Features of the Investigation of Conformational Transitions by the Fluorescence Method

Conformational transitions may be due to a number of external factors, such as temperature, solvent composition, or pH. The special features of the study of such transitions by fluorescence is that the factors of the external medium that are capable of changing the structure of the protein also exert a direct effect on the fluorescence of the aromatic amino acid residues in the protein. Thus the overall change in the fluorescence intensity in response to some external action is composed of this direct effect on the fluorescence of the chromophore and of an indirect effect due to a structural change of the protein. The contribution of these two effects to the overall change will vary from case to case.

Consider, for example, the pH dependence of the fluorescence intensity. Some tryptophan-containing proteins do not undergo structural changes in acid media, but their fluorescence is slightly quenched by COOH groups [Steiner and Edelhoch (1963*a*)]. The usual course of the pH dependence of fluorescence is disturbed in these proteins for which other methods have revealed conformational changes in this range of pH. In these cases the pH dependence depends both on the conformational changes of the protein and on the carboxyl quenching [Steiner and Edelhoch (1963*a*)]. Similarly, the character of the dependence of the fluorescence of proteins on pH in the alkaline region is determined by the effect of conformational transitions and the quenching effect due to ionization of the tyrosine hydroxyls [Frattali *et al.* (1965)].

Conformational transitions of proteins under the influence of temperature changes lead to a disturbance of the monotonic temperature dependence of the fluorescence intensity, and sometimes to the appearance of irreversibility or spectral shifts [Steiner and Edelhoch (1962*a*, 1963*a*); Gally and Edelman (1962, 1964)].

Figure 1.10 presents, as an example, the temperature dependence of the fluorescence properties of trypsin [see Turoverov *et al.* (1967*a*); Barenboim *et al.* (1968)].

The region of disturbance of the monotonic temperature dependence of the fluorescence intensity (from 49 to 62°C) corresponds well to the temperature region in which, as is indicated by results obtained by other methods, there is a conformational transformation of the trypsin macromolecules.

The dependence is entirely reversible in the temperature interval between 15 to 49°C and from 62 to 73°C. In the interval between 49 to 62°C,

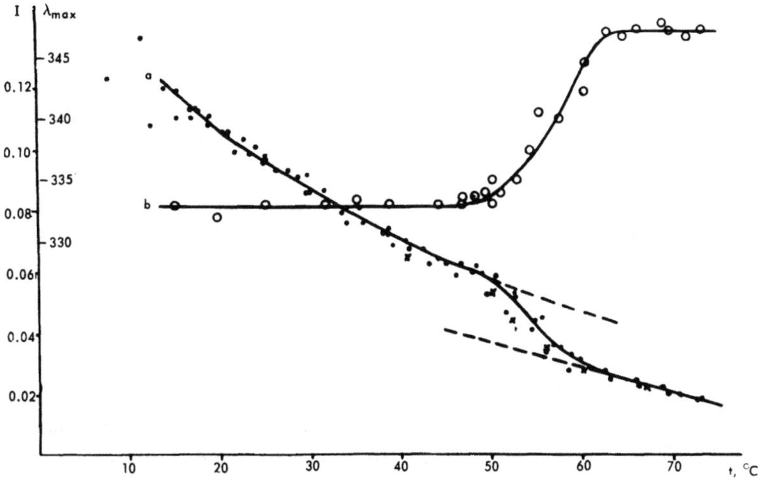

Fig. 1.10. Temperature dependence of the fluorescence properties of trypsin. (a) Temperature dependence of the fluorescence intensity (I) (λ_{fl} = 333 mμ); and (b) temperature dependence of the position of the maximum of the fluorescence spectrum (λ_{max}). Concentration: 0.005–0.02%; 0.15 M citrate-phosphate buffer (pH 3.4). λ_{exc} = 280 mμ.

there is reversibility only in the case of sufficiently slow cooling.

Together with an investigation of the temperature dependence of the fluorescence intensity, we measured the fluorescence spectra within the entire temperature interval (Fig. 1.10). In the temperature range from 15 to 49°C, the shape and position of the fluorescence spectrum remain unchanged (λ_{max} = 333 mμ). Upon further heating the spectrum is shifted in the long-wave direction. The spectral shift reaches a maximum (14 mμ) at 62°C (λ_{max} = 347 mμ). Further heating is not accompanied by any change in the position and shape of the spectrum.

These results, together with the disturbance of the monotonic temperature dependence of the fluorescence intensity, also are evidence of the presence of a conformational transition of trypsin in the temperature range from 49 to 62°C.

The long-wave shift of the spectrum is evidence that the tryptophan residues, which are in hydrophobic surroundings in the native macromolecule, become accessible to the solvent (pass into aqueous surroundings) as a result of the transition, which in turn is evidence of an uncoiling of the compact globular structure of the trypsin macromolecules. This corresponds to what is mentioned in the literature regarding the conformational transformation of trypsin during heating.

A monotonic decrease in the fluorescence intensity during heating will be observed in biopolymers that do not undergo conformational transformations under the influence of temperature changes. An illustration of this may be found in the temperature dependence of the fluorescence intensity of ribosomes (Fig. 1.11), for which, according to the data of sedimentation analysis conducted parallel with a measurement of the fluorescence [see Barenboim *et al.* (1967*c*)], there were no changes in the structure of the ribosome particles. The temperature dependence of the fluorescence intensity was entirely reversible to a temperature of approximately 52°C. At temperatures above 52°C, where there is an irreversible degradation of the ribosome particles, the monotonic temperature dependence of the fluorescence intensity is disturbed, and irreversibility appears (Fig. 1.11).

To explain the temperature dependences obtained for the fluorescence intensities of trypsin and ribosome, we must consider that a change in the temperature may influence the fluorescence intensity in the following manner: (1) through the direct influence of temperature on the fluorescence properties of tryptophan residues (a physical process of purely temperature quenching); and (2) through a change in the properties of the microenvironment of the tryptophan residues during conformational transitions.

The processes explaining the temperature quenching of fluorescence were discussed in the Introduction (Section 0.5). Gally and Edelman (1962), using simple transformations, obtained an equation from Eq. (0.7) which is cited below in a somewhat modified form:

$$\ln\left[\frac{d\left(-\frac{1}{q}\right)}{d\left(\frac{1}{T}\right)}\right]=\ln\left(\frac{k_3\Delta E}{k_1}\right)-\frac{\Delta E}{RT}\,. \tag{1.2}$$

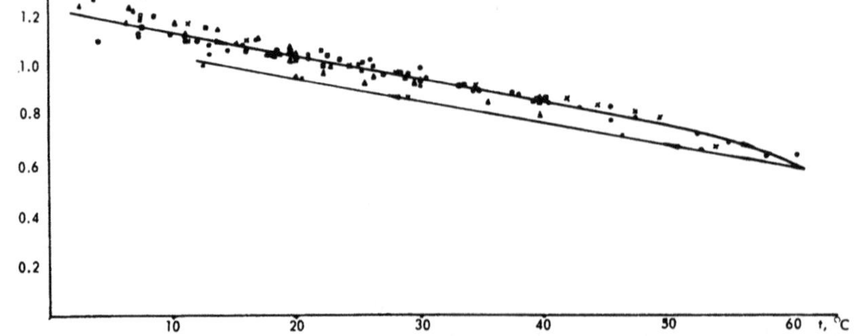

Fig. 1.11. Temperature dependence of the fluorescence intensity of the ribosomal 50S + 30S particles washed with NH_4Cl in 10^{-2} *M* phosphate buffer, 10^{-2} *M* $MgCl_2$. λ_{exc} = 296 mμ; λ_{reg} = 345 mμ.

The dependence of q upon T corresponds to purely temperature quenching of fluorescence when it obeys Eq. (1.2), i.e., there is a linear dependence of $[d(-1/q)/d(1/T)]$* upon $1/T$.

An analysis of the curves in Figs. 1.10 and 1.11 indicated that the temperature dependence of the fluorescence intensity obeys Eq. (1.2) for trypsin in the temperature range from 15 to 49°C and from 62 to 73°C, and for the ribosomes in the entire temperature range up to 52°C.†

On the basis of these results, it may be concluded that the decrease in the fluorescence intensity of trypsin during heating at temperatures between 15 to 49°C is due to purely temperature quenching of the fluorescence of the tryptophan residues of the native macromolecule. Upon further heating, there is a structural rearrangement of the trypsin macromolecules (T_{rearr} = 55°C), leading to a change in the properties of the microenvironment of the tryptophan residues and a change in their fluorescence properties – and thereby to a disturbance of the monotonicity of temperature quenching. At temperatures above 62°C, there is a purely temperature quenching of the fluorescence of the tryptophan residues of trypsin which has undergone a conformational transformation.

In the case of the ribosomes, the decrease in the fluorescence intensity in the entire temperature range up to 52°C is due to purely temperature quenching of the fluorescence of tryptophan residues.

The activation energy of temperature quenching ΔE, characterizing the electronic properties of the tryptophan residues, was found to be different for trypsin before and after the conformational transition, as well as for trypsin and the ribosomes (Table 1.5). This may be due to two factors: (1) a change in the fluorescence properties when tryptophan is inserted into the polypeptide chain (the influence of the neighboring amino acid residues and peptide groups); and (2) the influence on the fluorescence properties of the tryptophan residue exerted by its microenvironment in the protein macro-

*The quantity $[d(-1/q)/d(1/T)]$ is determined graphically from the dependence of $1/q$ on $1/T$.

†This fact seems unexpected, since the temperature dependence of the fluorescence intensity of proteins is determined by the superposition of temperature quenching of the fluorescence of the individual tryptophan residues (four in the case of trypsin and of the order of a hundred in the 50S + 30S ribosome particle), and therefore may not obey Eq. (1.2). The temperature dependence of the fluorescence intensity of protein will evidently obey Eq. (1.2) in three cases: (1) in the presence of only one tryptophan residue on the macromolecule; (2) in the presence of several tryptophan residues situated under identical or close conditions (this case may be realized in trypsin); and (3) in the presence of a number of tryptophan residues with complete averaging. According to the data of Gally and Edelman (1964), the temperature quenching of fluorescence satisfies Eq. (1.2) for a number of peptides containing tyrosine and (or) tryptophan, while among the group of proteins investigated (pepsin, human γ-globulin, egg albumin, bovine serum albumin, lysozyme), the equation is satisfied only for lysozyme.

Table 1.5 Fluorescence Characteristics and Activation Energies of the Temperature Quenching of Trypsin and Ribosomes

Preparation	Temperature region	λmax mμ	ΔE. (kcal/mole)
Trypsin	15–49°C	333	4.8
	62–73°C	347	8.8
Ribosomes	5–52°C	333	5.7

molecule (the polarity and viscosity of the medium, the presence of quenching groups close to the chromophore, the rigidity of the attachment of the chromophore, etc.).

The influence of these factors may vary both for the individual proteins, and for the same protein in different states.

The change in the fluorescence properties of tryptophan when it is inserted into a polypeptide chain may be considered as a particular case of the influence of substituents on the fluorescence properties of the indole ring. The mechanism of the influence of substituents is not known. Therefore, to explain the temperature dependence of the fluorescence intensity of proteins, it was necessary first to investigate the temperature quenching of the fluorescence of indole and its derivatives [see Turoverov *et al.* (1967*b*)].

A monotonic decrease in the quantum yield with increasing temperature was detected for all the investigated indole derivatives (Fig. 1.12). Table 1.6 presents the constants of temperature quenching of fluorescence ($\Delta E, k_2/k_1, k_3/k_1$) of derivatives calculated on the basis of the curves of Fig. 1.12 and Eqs. (1.2 and 0.7). The activation energy of temperature quenching ΔE correlated with the quantum yield q in the series of indole derivatives (Fig. 1.13).

The results obtained are evidence that the addition of various substituents to the indole ring and, as a particular case of this, insertion of a tryptophan residue into the polypeptide chain, influences the course of temperature quenching. This is evidence that the electronic state of the neighboring animo acids and peptide groups (which may differ both in the case of different proteins and in the case of different states of the same protein) is capable of determining to some degree the difference in the course of temperature quenching which is observed experimentally between trypsin and the ribosomes and between trypsin in its native and denatured states. Nothing more concrete can be said about the nature of this influence since there is generally no single opinion on the mechanism of the influence of substituents on the fluorescence properties of the indole ring.

Now we have begun experiments on a systematic study of the influence of external quenchers and the polarity of the medium (factors of the microenvironment of the tryptophan residues in the protein) on the course of temperature quenching of indole derivatives. These investigations, on the one hand, are outlining some approaches to a solution of the mechanism of the influence of substituents on the fluorescence properties of the indole ring, and, on the other hand, permit us to hope that we will be able to solve the opposite problem – how to judge the properties of the microenvironment of the tryptophan in the protein macromolecule according to the temperature dependence of the fluorescence. The latter is important in view of the special role of tryptophan residues in the active centers of proteins and in the organization of supermolecular structures which is being discussed in the literature.

Studies of the Kinetics of Conformational Transitions and Determination of Thermodynamic Characteristics

The time dependence of the fluorescence intensity in the course of a conformational transition has been studied by several authors [Cowgill (1964*a*); Steiner *et al.* (1964); Edelhoch *et al.* (1965)], and has always been found to satisfy closely a first-order rate equation (Fig. 1.14).

$$\frac{1}{t}\ln\frac{I-I_\infty}{I_0-I_\infty}=k. \tag{1.3}$$

Table 1.6. Constants of Temperature Quenching of the Fluorescence of Tyrosine and Indole Derivatives in Aqueous Solution

	Compound	ΔE (kcal/mole)	$\frac{k_3}{k_1}$	$\frac{k_2}{k_1}$
1.	Indole	9.9	$2.8 \cdot 10^7$	0.25
2.	Indolyl-3-propionic acid	11.8	$4.4 \cdot 10^8$	0.37
3.	Indolyl-3-butyric acid	11.8	$3.4 \cdot 10^8$	0.24
4.	D, L-Tryptophan*	8.1	$1.8 \cdot 10^6$	1.6
		(8.1)	$(2.2 \cdot 10^6)$	(1.3)
5.	N-acetyltryptophan	9.4	$1.7 \cdot 10^7$	0.26
6.	Monoglyceride of indolyl-3-propionic acid	6.6	$2.7 \cdot 10^5$	1.26
7.	Glycyltryptophan	8.2	$1.7 \cdot 10^7$	1.16
8.	Tyrosine*	7.1	$4.3 \cdot 10^5$	0.9

*The temperature quenching constants obtained by Gally and Edelman (1962) are cited for tryptophan (in parentheses) and tyrosine.

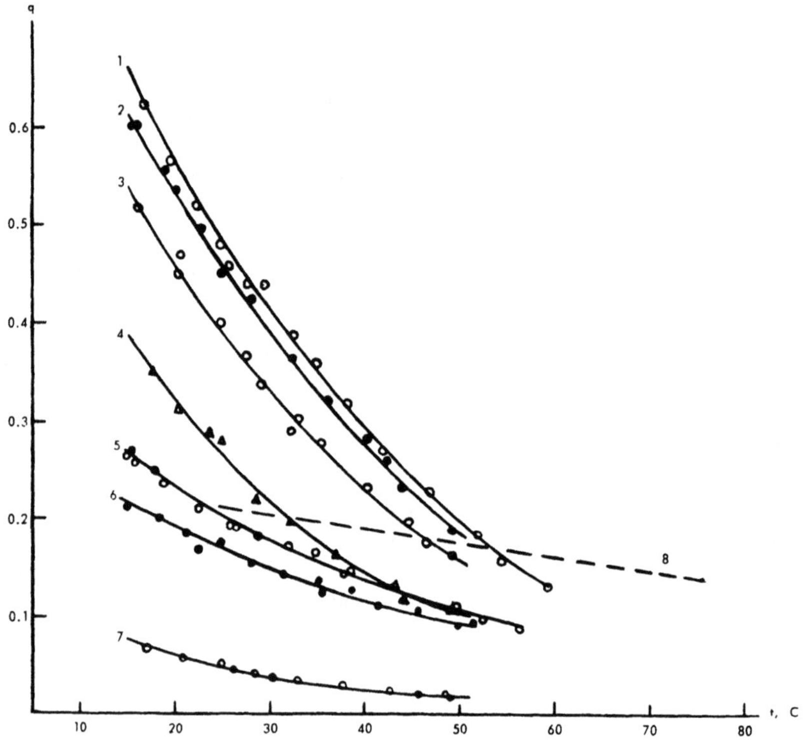

Fig. 1.12. Temperature dependence of the fluorescence quantum yields of tyrosine and indole and its derivatives: (1) indolyl-3-butyric acid; (2) indolyl-3-propionic acid; (3) indole; (4) N-acetyltryptophan; (5) tryptophan; (6) monoglyceride of indolyl-3-propionic acid; (7) glycyltryptophan; and (8) tyrosine. Curves 1–7 were obtained by the present authors together with V. A. Sovolenko. Curve 8 comes from the work of Gally and Edelman (1962). The course of the temperature quenching for tryptophan corresponds to that given by Gally and Edelman (1962).

In Eq. (1.3) I_0, I, and I_∞ are, respectively, the relative fluorescence intensities at the initial moment of time, at some time t and at equilibrium, and k is the first-order rate constant.

The values of k obtained in this way are in good agreement with those determined on the basis of optical rotation, differential absorption spectrum, and so on. Fig. 1.15 shows the variation of k for the denaturation of pepsinogen, determined by various methods, as a function of the urea concentration.

The temperature variation of k given by the Arrhenius equation

$$\frac{d \ln k}{dT} = \frac{E^*}{RT^2}, \tag{1.4}$$

(where T is the absolute temperature, E is the activation energy, and R is the gas constant), allows a determination of the thermodynamic parameters of the conformational transition and of the activation energy. By integrating (1.4) we obtain

$$\ln k = -\frac{E^*}{RT} + c, \tag{1.5}$$

so that E can be found from the slope of a plot of $\ln k$ versus $1/T$.

The free energy of activation ΔF^* is calculated from

$$k = \frac{k'T}{h} e^{-\frac{\Delta F^*}{RT}}, \tag{1.6}$$

where h is Planck's constant and k is Boltzmann's constant. Since

$$\Delta F^* = \Delta H^* - T\Delta S^*, \tag{1.7}$$

where ΔH^* is the enthalpy and ΔS^* the entropy of activation, and since

$$\Delta H^* = E^* - RT, \tag{1.8}$$

it is easy to calculate ΔH^* and ΔS^* as well.

Edelhoch *et al.* (1965) determined in this way the thermodynamic parameters for the denaturation of pepsinogen with urea (Table 1.7).

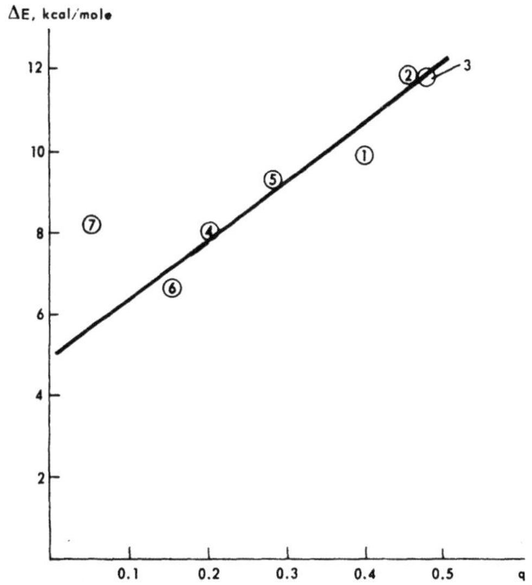

Fig. 1.13. Correlation relationship between the quantum yield ($q_{25°C}$) and the activation energy of temperature quenching (ΔE) for indole derivatives in aqueous solution. The numbers of the points correspond to the numbers of the indole derivatives in Table 1.6.

dried and crystalline proteins were quenched less readily than proteins in solution, evidently because the tryptophan residues in them were less accessible to oxygen.

Comparison of the results of quenching with potassium iodide and oxygen shows how important it is to select the molecular size of the quenching substance for a determination of the "topography" of the aromatic residues: the optimum quenching substance should penetrate just so far (and no further) as the molecules of water. In such a case the short-wave component due to the fluorescence of tryptophan residues in hydrophobic surroundings should appear more sharply during the quenching. Moreover, when an optimum quenching substance is used, denaturation of a protein which increases the accessibility of tryptophan residues to the solvent, should change the course of the dependence of I_0/I on C (where I and I_0 are, respectively, the fluorescence intensity in the presence of the quenching substance in concentration C and in the absence of that substance), owing to the change in the quantum yield of tryptophan residues passing from hydrophobic to hydrophilic surroundings. This phenomenon may serve as a kind of an indicator of a conformational transition, though no such studies have as yet been done with the classical luminescence-quenching substances.

Steiner *et al.* (1964) proposed a slightly different method of estimating the solvent-accessibility of tryptophan residues based on the dependence of the fluorescence intensity on solvent composition. It was found that the relative fluorescence intensity of acetyltryptophanamide (ATA) (a compound convenient for modeling the fluorescence properties of tryptophan in proteins because its amino and carboxyl groups are blocked) depends directly on the volume fraction of various substances added to the buffer at pH 7 (Fig. 1.8). The effect of one such additive, propylene glycol, on the fluorescence intensity of proteins was then investigated. The choice of propylene glycol was motivated by the fact that in small concentration in aqueous solution it exerts very little effect on the structure of the protein macromolecule [Herskovits and Laskowski (1962)]. This part of the work showed that the dependence of the relative intensity on the volume fraction of propylene glycol was the same for proteins with disturbed secondary and tertiary structure and for ATA (Fig. 1.16). In this context it should be noted that the variation of the fluorescence of ATA with the content of propylene glycol is unaffected by the presence of a denaturing agent such as urea (Fig. 1.16). Thus we are led to the conclusion that the influence of propylene glycol on the fluorescence of tryptophan residues does not depend on the primary structure of the protein.

The slope dI/dp in Fig. 1.16 proved to be a specific feature of native proteins, being lower for these than for ATA. Denaturation of the proteins with urea increased the slope, i.e., enhanced the solvent-dependence of the fluorescence of tryptophan residues. These experiments led the authors to

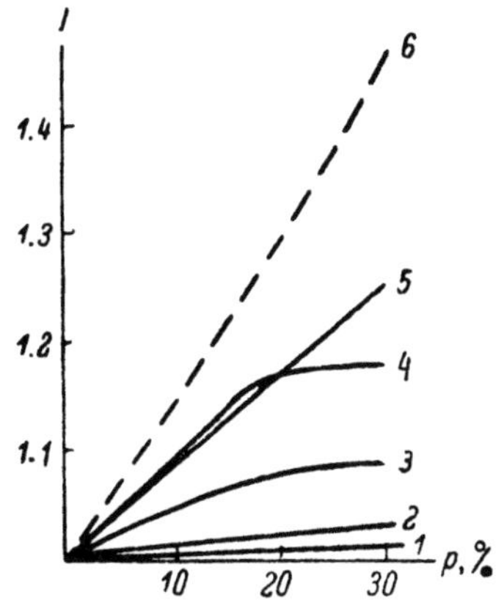

Fig. 1.16. Variation of the relative fluorescence intensity (I) of various proteins and of acetyltryptophanamide with the volume fraction (p) of propylene glycol in the solution [Steiner *et al.* (1964)]. (1) Pepsinogen; (2) bovine serum albumin; (3) pepsin; (4) lysozyme; (5) soy bean trypsin inhibitor; (6) ATA in aqueous solution in 7 M urea, bovine serum albumin in 8 M urea, and thyroglobulin treated with trypsin in 7 M urea (the curves approximately coincide).

conclude that the effect of neutral solvents on the fluorescence of tryptophan residues in proteins depends wholly on the nature of the secondary and tertiary structure, the slope of the curves being determined by the accessibility of these residues to the solvent. From this point of view the tryptophan residues in pepsinogen, albumin, and pepsin are less accessible than those in lysozyme or soybean trypsin inhibitor (Fig. 1.16). Denaturation by urea thus makes the protein structure more open and penetrable by the solvent.

The fluorescence intensity of a protein like the soybean trypsin inhibitor increases almost linearly with increasing urea concentration in the solution, which corresponds to the fact that urea up to a concentration of 9 M does not really disturb the structure of this protein [Edelhoch and Steiner (1963)]. Evidently this linearity shows that in this concentration range the additive does not change the protein structure. Conversely, the nonlinearity of the dependence of I on the propylene glycol concentration in pepsin and lysozyme solutions (Fig. 1.16) indicates that above a certain concentration propylene glycol affects the structure of these proteins.

On the basis of the experiments carried out by Steiner *et al.* (1964) it seems that in a number of cases the changes in dI/dp in the course of a conformational transition may reflect the structural rearrangement more directly than changes in I itself. Thus, the changes in the fluorescence intensity of proteins acted upon by urea are very complex (Fig. 1.18) and difficult to interpret. The effect of urea on protein fluorescence is determined not only by the structural changes in the macromolecules but also by the direct influence of urea on the fluorescence of tryptophan residues (Fig. 1.8) and by the changes in the tryptophan surroundings brought about by interaction of urea with various groups of the polypeptide chain. It is hardly possible to isolate the effect of the structural rearrangement from the overall picture. The advantage of dI/dp as a criterion of structural changes in the

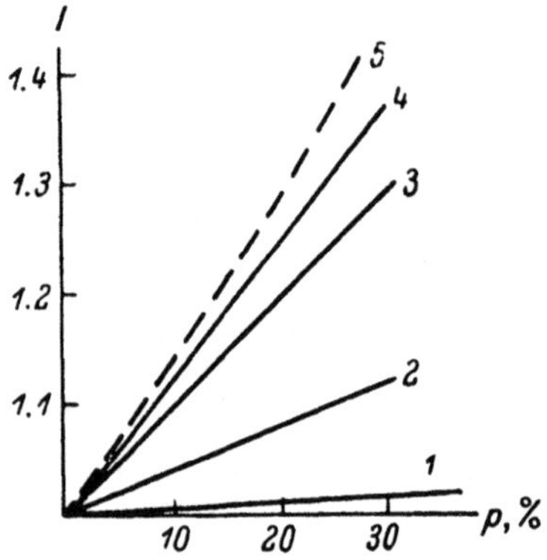

Fig. 1.17. Variation of the relative fluorescence intensity of thyroglobin with the volume fraction of propylene glycol in the solution [Steiner *et al.* (1964)]. (1) In the absence of urea; (2) in 3 *M* urea; (3) in 7 *M* urea; (4) in 0.02 *M* solution of sodium dodecyl sulfate; (5) in 9 *M* urea (coincides with the dependence for ATA).

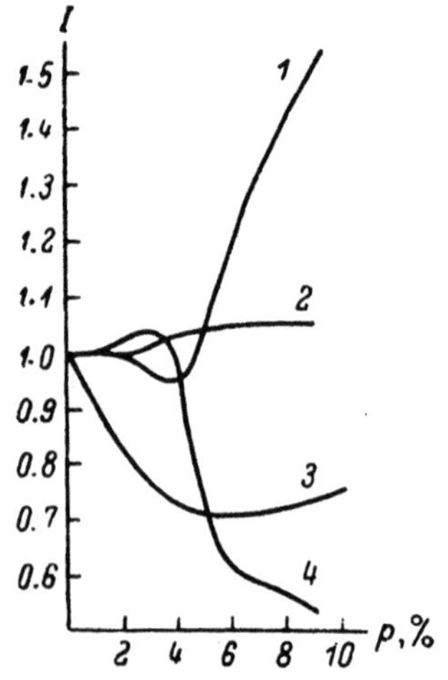

Fig. 1.18. Variation of the relative fluorescence of proteins with the volume fraction of urea in the solution [Steiner *et al.* (1964)]. (1) Thyroglobulin; (2) pepsinogen; (3) lysozyme; (4) human serum albumin.

macromolecule is that the influence of neutral additives is wholly (or almost wholly) determined by the secondary and tertiary structure of the protein, and is independent of the presence of a denaturing agent (e.g., urea).

★ ★ ★

The results reviewed in this section indicate that protein fluorescence is extremely sensitive to structural changes in the macromolecule. The use of the fluorescence of a protein for the study of conformational transitions makes it possible to: (1) detect conformational transitions and determine the conditions (temperature, pH, etc.) under which they take place; (2) assess the reversibility of such transitions; (3) study transition kinetics; (4) determine the thermodynamic parameters of the activation of conformational transitions; and (5) estimate the solvent-accessibility of aromatic amino acid residues in the protein, i.e., the hydrophobicity or hydrophilicity of the surroundings, and follow the changes in accessibility during conformational transitions.

In certain cases the changes in the fluorescence intensity of proteins may be followed directly in an enzyme–substrate interaction [Burshtein (1961); Sturtevant (1962); Burshtein and Suslova (1964)].

The fluorescence method allows one to work with very small amounts of the substance. Perlman (1964) used 1% pepsinogen solution to study optical rotation and 0.025% solution to study the fluorescence. If necessary, the protein concentration may be reduced even further, and in addition the actual volume of the solutions may be very small. Thus, conformational transitions in proteins can be studied by the fluorescence method with amounts a few thousand times smaller than those required for corresponding studies by other methods (especially by optical rotation). The ease of fluorescence measurements and the possibility of working with low concentrations make the fluorescence method convenient for studying the kinetics of denaturation and renaturation of proteins. Finally, the great advantage of the fluorescence method is that it allows a study of structural rearrangements of proteins directly in living cells and tissues; for example, this was the method used by Shtrankfel'd (1964*c*) to study structural changes occurring in muscle tissue with rising temperature.

1.5. POLARIZATION OF THE FLUORESCENCE OF AROMATIC AMINO ACIDS AND PROTEINS

One of the methods of investigating the properties of protein macromolecules in solution is studying the polarization of the fluorescence of their complexes with dyes [Steiner and Edelhoch, (1962*b*)], in which the fluorescent moiety of interest is the dye. We shall now consider a relatively less well known phenomenon, namely polarization of the intrinsic UV fluorescence of proteins. This phenomenon may be used to study the electronic structure of chromophores, to determine the molecular volumes of low-molecular proteins, to identify certain proteins, to observe structural changes, and to study energy migration processes

Study of the Electronic Structure of Chromophores

To study the nature of the oscillators of aromatic amino acid residues in proteins we need to examine the polarization spectra of the free amino acid. Not stopping to interpret these spectra in detail, we shall here merely summarize the main results concerning the electron structure of amino acids obtained from the polarization of their fluorescence.

Indole and Tryptophan. The polarization absorption spectra of indole and tryptophan and propylene glycol at -70°C (shown in Fig. 1.19) were interpreted

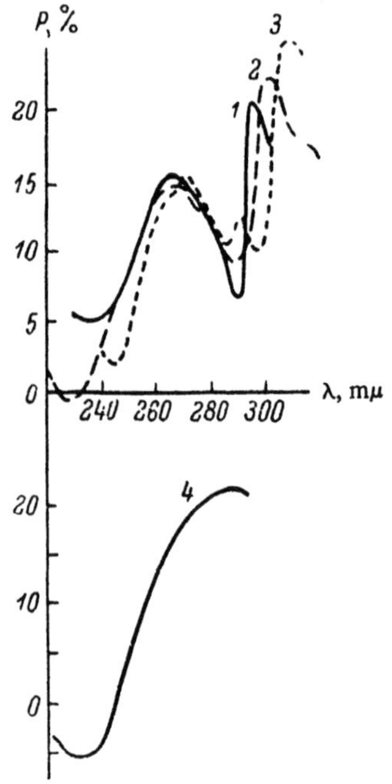

Fig. 1.19. Polarization spectra of (1) indole, (2) tryptophan, (3) N-glycyltryptophan, and (4) tyrosine in propylene glycol at -70°C [Weber (1960*a*)]. *P* is the degree of polarization.

by Weber (1960*a*, 1961*a*) using the postulate that the long-wave absorption band of these amino acids is formed by two electronic transitions corresponding to two oscillators, ^1L_a and ^1L_b (Fig. 1.16). The oscillators lie in the plane of the molecule and are oriented at nearly right angles to each other. The absorption band of ^1L_b lies within the absorption band of ^1L_a. The oscillator ^1L_b does not itself emit, but transfers the excitation energy to ^1L_a which is responsible for the fluorescence; this is the cause of the "fall" in the polarization spectrum in the region of 270–300 mμ where both oscillators absorb. Maximum depolarization of the fluorescence occurs when the excitation wavelength is 289 mμ at which the contribution of oscillator ^1L_b to the absorption is at its highest. Weber's postulate of two oscillators participating in the absorption spectrum of tryptophan was later confirmed by investigations of the polarization of tryptophan fluorescence in films of polyvinyl alcohol [Sevchenko *et al.* (1963 *a*); Chernitskii *et al.* (1963); Chernitskii and Konev (1965)].

On the other hand, the idea that the fluorescence spectrum is determined solely by oscillator ^1L_a proved to be wrong: ^1L_b can itself emit in the short-wave region of the fluorescence spectrum [Zimmermann and Joop (1961); Chernitskii *et al.* (1963)]. This conclusion is reached on the basis of the observation that the degree of polarization falls off in the short-wave end of the fluorescence spectrum [Zimmermann and Joop (1961)] and because the polarization absorption spectrum changes its form when the fluorescence in the short-wave region is recorded [Chernitskii *et al.* (1963)]. Thus, in films of polyvinyl alcohol the oscillator ^1L_b passes most of the energy to ^1L_a but also emits to some extent [Chernitskii *et al.* (1963)].

Studies of the polarization spectra of indole and tryptophan in isotropic and anisotropic film of polyvinyl alcohol made it possible to determine not only the orientation of the oscillators with respect to the direction of the stretching of the film [Sevchenko *et al.* (1963*a*)], but also relative to the axis of the molecule [Chernitskii and Konev (1965)]. It was found that ^1L_b is directed parallel to, and ^1L_a forms an angle smaller than 90° with the long axis.

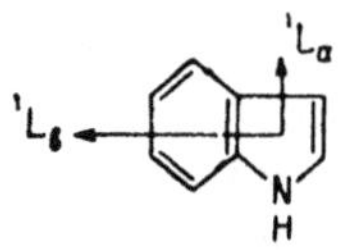

In drawn films of polyvinyl alcohol oscillator ^1L_a orients itself in the direction of drawing.

Less work has been done on the oscillatory nature of the phosphorescence of indole and tryptophan, Chernitskii and Konev (1965) held the view that the phosphorescence is determined by the transition $A \leftarrow {}^3L_a$. One of their arguments in support of this hypothesis is the coincidence of the shape of the fluorescence and phosphorescence spectra of crystalline amylase [Bobrovich and Konev (1964)]. However, as already mentioned, even the statement that the fluorescence spectrum of crystalline amylase is determined by the transition $A \leftarrow {}^1L_a$ is open to some doubt. According to Chernitskii and Konev (1965), the phosphorescence oscillator is directed almost perpendicularly to the plane of the molecule and has a component parallel to the long axis of the molecule, i.e., along ^1L_b.

Phenol and Tyrosine. Solutions of phenol and tyrosine give the same polarization spectrum (Fig. 1.19). The spectra were analyzed by Weber (1960*a*), who came to the conclusion that they are determined by two transitions with maxima at 235 and 275 mμ, the oscillators of which lie in the plane of the molecule and are almost perpendicular to each other.

This study of polarization spectra has made it possible to analyze the main aspects of the spectra of aromatic amino acids. In our opinion a large gap in the investigations of the polarization spectra of indole and tryptophan fluorescence is the absence of any systematic work on the influence exerted on these spectra by the polarity of the medium. Such investigations would allow an assessment of the contributions of oscillators ^1L_a and ^1L_b to the fluorescence spectra in media of various polarity. This is important for an understanding of the fluorescence properties of tryptophan residues in proteins in general, and in particular for an understanding of the polarization spectra of tryptophan-containing proteins.

The close similarity between the polarization spectra of indole and tryptophan fluorescence indicates that substitution in the indole ring does not change its oscillatory properties. This in turn suggests that such properties remain essentially unchanged when tryptophan is included in a polypeptide chain.

Polarization Spectra of Proteins

The polarization spectra of globular proteins were investigated by Weber (1960*a*, 1961*a*), who found that the spectra of the tryptophan-containing proteins on the whole resemble those of free tryptophan (and indole) in solution, though some general and some very specific differences could be observed (Fig. 1.20). In comparison with the spectra of indole and tryptophan, most tryptophan-containing proteins showed an additional narrow peak with a maximum at 290 mμ. This appeared most strongly for lysozyme, less so for albumins, and not at all in egg albumin. The presence of the same additional peak in the polarization spectrum of N-glycyltryptophan led Weber to propose that it is associated with the peptide linkage.

A general feature of the polarization spectra of proteins is the large depolarization of the fluorescence in comparison with the free amino acids. This is particularly strong in the region of 270 mμ. The ratio of the degrees of

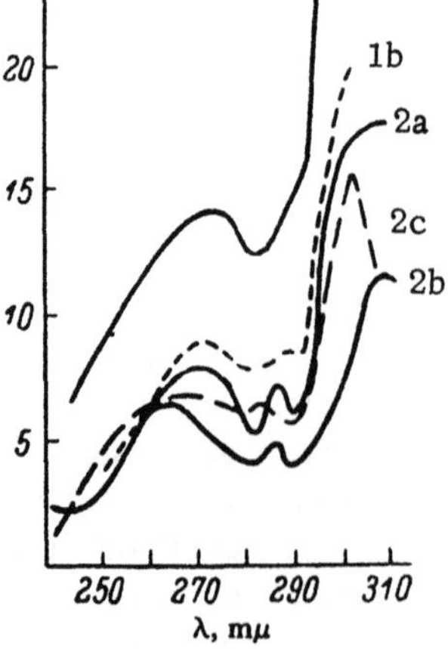

Fig. 1.20. Polarizaiton spectra of proteins. (1) Excitation with polarized light [Bobrovich and Konev (1965)]: (a) human serum albumin; (b) chymotrypsinogen; (2) excitation with unpolarized light [Weber (1960*b*)]: (a) human serum albumin; (b) human serum albumin denatured with 8 *M* urea; (c) pepsin. (*P* is the degree of polarization.)

polarization at 305 and 270 mμ (P_{305}/P_{270}) proved to be a parameter characteristic of individual proteins, being a function of the protein structure [Weber (1960*a*, 1961*a*)]. For indole, tryptophan, and N-glycyltryptophan this ratio is 1.4–1.7, for proteins in aqueous solution at room temperature 1.8–2.5 and, for proteins in a 1:1 mixture of water and propylene glycol at -70°C, 2.2–3.0.

No unequivocal interpretation of the polarization spectra of tryptophan-containing globular proteins can as yet be given. In Weber's opinion, the difference between the spectra of these proteins and of free tryptophan in solution is due to the inclusion of the tryptophan into the polypeptide chain, and to special features of the microenvironment of the tryptophan residues. Not denying a possible influence of the protein structure on the oscillatory nature of the tryptophan residues, Konev *et al.* (1965) considered that the depolarization of protein fluorescence is caused mainly by intertryptophan resonance energy migration processes. In our own opinion, the effect of structure could be very important because the screening of tryptophan residues from the solvent must be taken into account. Such screening can modify the polarity of the surrounding medium and so change the contributions of ^1L_a and ^1L_b to the fluorescence and absorption spectra. The reduction in solvent polarity may be the cause of the increase of(P_{305}/P_{270})to 1.4–1.9 on denaturation of proteins with urea observed by Weber (1960*a*, 1961*a*).

An effect of protein structure on luminescence polarization was found in an investigation of the fibrous proteins [Konev and Katibnikov (1961)], whose structure changes on extension. The luminescence of the randomly oriented keratin fibers of wool, and that of silk fibrion, is strongly depolarized. Extension of the fibers gives rise to a rapid increase of the degree of polarization of the fluorescence, so that *P* is sensitive to the structure of proteins belonging to this group and may be used to follow conformational changes.

The polarization spectra of proteins not containing tryptophan are similar in form to the spectra of phenol and tyrosine, except that in the former case the absolute degree of polarization is smaller over the entire spectrum [Weber (1960*a*)].

★ ★ ★

Although it can be seen from the foregoing remarks that the interpretation of the polarization spectra of proteins is still rather uncertain, the fluorescence polarization spectra are individual and highly characteristic for many tryptophan-containing proteins, being very sensitive to structural factors. It is possible that correlations will eventually be found between the differences in polarization spectra and structural peculiarities of the proteins.

Polarization of the fluorescence of protein–dye complexes is utilized to determine the size of protein macromolecules. It is therefore natural to ask whether the intrinsic UV fluorescence of proteins could not be used for this purpose. The molecular volume ν of a fluorescing molecule is determined on the basis of Perrin's formula (0.5) from the experimental dependence of $1/P$ on T/η if the lifetime τ of the excited state is known. The overwhelming majority of proteins are so large that their rotational relaxation times are very considerably larger than the lifetime of the excited state of aromatic amino acid residues, even in media of low viscosity, e.g., in aqueous solutions at room temperature. The degree of polarization is practically independent of viscosity and temperature and cannot provide any information about the volume of the macromolecule.

On the other hand, polarization of UV fluorescence makes it possible to determine the molecular volumes of tyrosine and tryptophan-containing polypeptides with molecular weights not exceeding a few thousand [Weber (1960*a*)].

The study of fluorescence polarization is also informative as regards energy migration processes. Since these processes were already treated in the preceding sections, we shall only note at this point that most of the data on the resonance migration of the excitation energy between aromatic amino acid residues in proteins were obtained by the study of fluorescence polarization.

1.6. ULTRAVIOLET FLUORESCENCE OF CELLS

These studies are of relatively recent origin. UV fluorescence of cells was first detected by E. M. Brumberg (1956) using a special microscope with an optical system transparent to ultraviolet. The cells could thus be photographed by the light of their own UV fluorescence. The possibilities of studying the phenomenon were further extended by the application of photoelectronics [Agroskin and Korolev (1961); Barenboim *et al.* (1962); Agroskin *et al.* (1964)]. It was shown that UV fluorescence microscopy can provide valuable information on various processes of life activity taking place in animal and plant cells in normal and pathological states [see Meisel' *et al.* (1961); Troshin and Barenboim (1963); I. Ya. Barskii (1964); E. M. Brumberg *et al.* (1964); Chernogryadskaya *et al.* (1964*b*, 1964*c*)].

General Characteristics of the UV Fluorescence of Cells

On excitation in the region of 250–300 mμ (λ_{max} = 280 mμ in the excitation spectrum) most of the normal and pathological cells and tissues in man, animals [see Chernogryadskaya *et al.* (1964*c*)], and plants investigated [I. Ya. Barskii *et al.* (1959)] exhibit UV fluorescence. The same is true of bacteria [Agroskin and Barskii (1961); E. M. Brumberg and Barskii (1961); Pomoshchnikova and Agroskin (1961)] and, according to our results, of bacteriophages.

The fluorescence spectra of most cells and tissues have a single band in the long-wave UV region, with a maximum at 330–350 mμ, which corresponds in its position to the maxima of tryptophan and tryptophan-containing proteins

(Fig. 1.21). The exact position of the fluorescence maximum is however slightly different in each case.

The fluorescence intensity depends on the object and on its functional state. According to Chernogryadskaya *et al.* (1964*c*), the brightest fluorescence is that of muscles and cells of epithelial tissues while the intensity from nerve fibres and blood cells is lower. Some cells do not fluoresce, for example, the reticulocytes and erythrocytes of peripheral blood [E. M. Brumberg *et al.* (1960*b*)].

The most accurate information can be obtained if one can determine exactly which cell structures, and which molecules of these structures, are responsible for the fluorescence, and these problem have naturally attracted special attention of investigators.

Fluorescence of Individual Cell Structures

The brightest fluorescence in cells comes from the cytoplasm [E. M. Brumberg *et al.* (1961*b*); Chernogryadskaya *et al.* (1964*c*)], although weak fluorescence can also be observed from nuclei both within the cell [E. M. Brumberg and Barskii (1961); V. E. Barskii *et al.* (1962); Khrushchev (1963)] and outside it [Barenboim *et al.* (1963*a*); Konev *et al.* (1963*b*); Konev and Bobrovich (1965)]. E. M. Brumberg and Barskii (1960, 1961) were able to record the fluorescence of chromosomes in the nuclei of cells from the salivary glands of chironomid larvae, but no chromosomal emission could be detected in the cells of other organisms [E. M. Brumberg *et al.* (1961*c*); Chernogryadskaya and Shudel' (1962*a*)]. The UV fluorescence of nucleoli is much stronger in comparison with that of nuclei [Khrushchev (1963); E. M. Brumberg *et al.* (1964)]. Within the cytoplasm, the mitochondria have the most intense fluorescence (Fig. 1.22), as was shown both on whole cells of insects, amphibia,

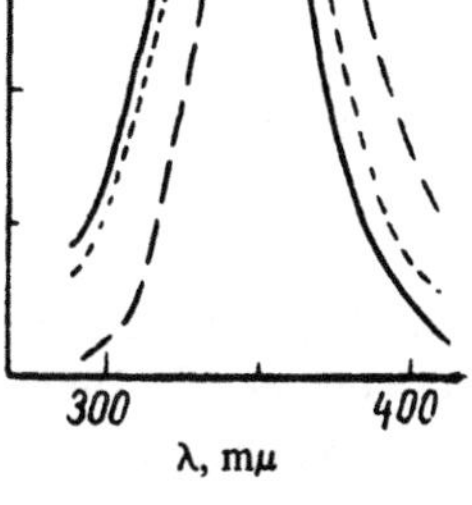

Fig. 1.21. Fluorescence spectra of (1) leucocytes, (2) serum albumin, and (3) tryptophan. *I* is the fluorescence intensity.

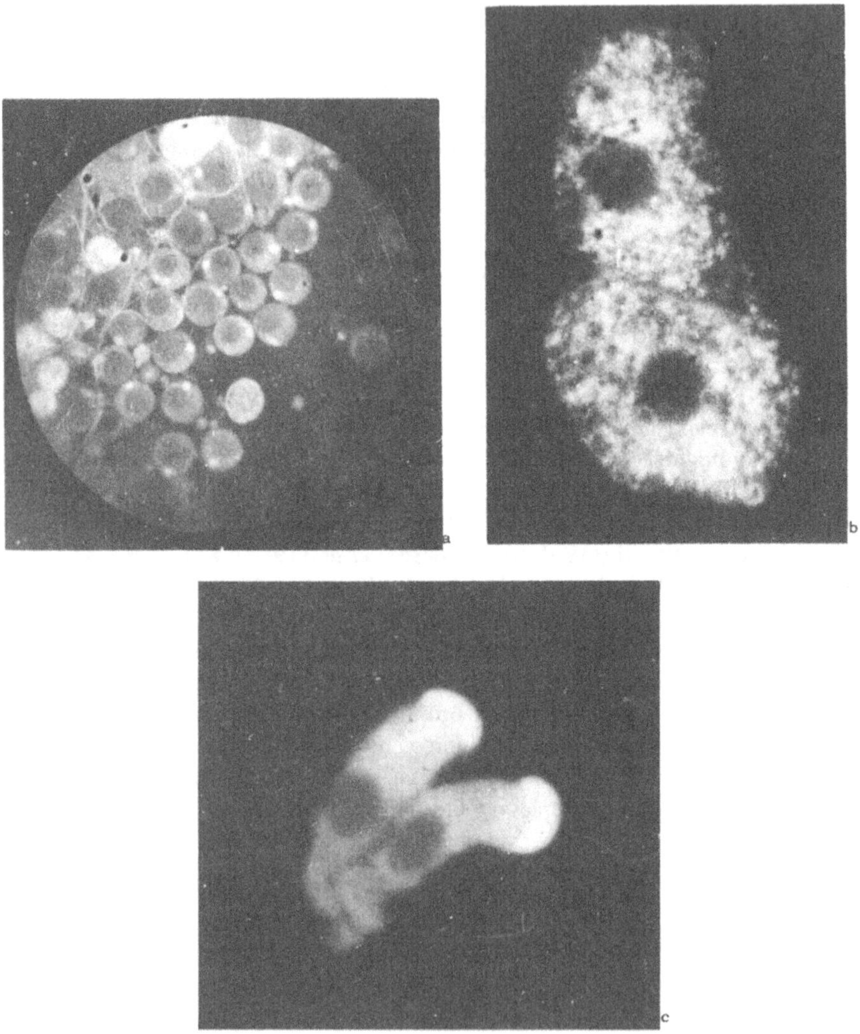

Fig. 1.22. Fluorescence of mitochondria in various cells. (a) Grasshopper spermatoblasts; bright fluorescence from mitochondrial masses (Nebenkern); (b) axolotl liver cells; the brightly fluorescent granules are mitochondria; and (c) epithelial cells of axolotl small intestine; the brightest fluorescence comes from parts of the cytoplasm above the nuclei where most of the mitochondria are located, and from the periphery of the cells.

and mammals [I. Ya. Barskii *et al.* (1962); Chernogryadskaya and Shudel' (1962*b*); E. M. Brumberg *et al.* (1963*a*); Chernogryadskaya *et al.* 1964*d*; Chernogryadskaya *et al.* (1964*b*)], and on cell organelles [Barenboim *et al.*

(1963*a*); Konev and Lyskova (1965*a*)]. The fluorescence of ribosomes isolated from *E. coli* has been discovered and investigated by the present authors. Some finer details of the localization of emission within the cell are given in the review by Chernogryadskaya *et al.* (1964*c*).

Substances Responsible for Cell Fluorescence

This problem has been the subject of lively discussion for a number of years, which is not surprising if we take into account the extreme chemical complexity of the living cell.

The fluorescence has been ascribed to proteins [E. Brumberg *et al.* (1958); E. M. Brumberg *et al.* (1963*b*)], nucleic acids [E. M. Brumberg and Barskii (1960, 1961)], free nucleotides [E. M. Brumberg *et al.* (1960*b*); E. M. Brumberg *et al.* (1960*a*); E. M. Brumberg *et al.* (1961*a*)] and nucleotides in complexes with cell structures ensuring rigid fixation and hence considerable fluorescence of the chromophore [E. M. Brumberg *et al.* (1961*b*)], nucleotide-type coenzymes [Chernogryadskaya and Shudel' (1962*a*)], and ubiquinones [V. E. Barskii *et al.* (1962); E. M. Brumberg *et al.* (1963*a*)]. In many of these studies the fluorescing substances were determined by the change in the fluorescence when individual substances were extracted from the cells. Even neglecting the fact that the nature of the emission from living cells was in this case studied by the emission of dead cells, the condition during such an extraction could have affected strongly the fluorescence quantum yield owing to changes in the pH of the medium, temperature, conformation of the biopolymers, interactions of the chromophore with organic and inorganic quenching groups, and so on. In spite of this, the possible influence of these factors, whose action on fluorescence is considered in detail in the preceding sections of this chapter, was not taken into account in the analysis of the results. Therefore the most reliable data on the substances responsible for cell fluorescence were obtained by spectral methods: Agroskin and Barskii (1961) showed that the excitation and fluorescence spectra of cells coincide with the analogous spectra of tryptophan-containing proteins. Such coincidence of the fluorescence spectra was found by Barenboim *et al.* (1961). Konev *et al.* (1963*a*) and Konev and Bobrovich (1965) showed that the fluorescence polarization spectra of mitochondria and nuclei fractions isolated from the liver cells of the white rat resemble, both in form and in absolute polarization, the polarization spectra of tryptophan-containing proteins. The idea that most of the cell fluorescence is due to proteins is today widely accepted [E. M. Brumberg *et al.* (1964)]. The absence of a nucleic acid component in the UV emission spectrum of cells is in full agreement with the fact that nucleic acids do not fluoresce at neutral pH and on excitation in the region of normal absorption (see Section 2.1).

Since mitochondria have the most intense fluorescence of all cell structures it could be postulated that the cell emission is due to the fluorescence of mitochondrial proteins. The connection between the UV fluorescence and the energy metabolism of the cell has been studied by a number of authors [Shudel' *et al.* (1964); Shudel' (1964*b*), (1965*a*); E. M. Brumberg and I. E. Brumberg (1964)]. The use of various metabolic poisons inhibiting electron transfer along the respiratory chain and displacing the redox level of respiratory enzymes in the direction of oxidation or reduction led to the conclusion that the fluorescence is due mainly to the proteins of the respiratory enzymes of mitochondria. According to a rough estimate given by E. M. Brumberg *et al.* (1964), 70–80% of the UV emission of Ehrlich ascites tumor cells is due to the tryptophan of the mitochondrial respiratory enzyme proteins. On the basis of their study of the fluorescence polarization spectra of mitochondria, Konev and Bobrovich (1965) concluded that the fluorescence comes exclusively from tryptophan residues. In the spectral excitation region of 250–296 mμ the spectra were independent of the excitation wavelength, which indicated the involvement of only one emission center in the cell emission; in their form the spectra coincided with the analogous spectra of tryptophan residues in proteins.*

Since all the proteins investigated that contain aromatic amino acid residues exhibit fluorescence (Section 1.2), why is it that in the cell the emission comes largely from the mitochondrial proteins? According to E. M. Brumberg *et al.* (1964), the reason is that in the cell the protein part of nucleoproteins does not fluoresce, and the various enzymes bound with quenching substrates or coenzymes fluoresce only weakly. On the other hand, combination of proteins with lipids favors an increase of the fluorescence quantum yield. This was demonstrated experimentally on the bright fluorescence from the myelin sheath of nerve fibers and the parts of the folded membrane of the outer surface of small intestine epithelium where the lipoprotein in membranes are packed most closely.

It is known that in mitochondria the large group of lipoproteins is composed of complexes formed between lipids and the protein components of the electron-transfer chain, and it is these proteins that are believed to be responsible for the UV fluorescence of the cell. If this is true, then other lipoprotein complexes entering into the composition of numerous cell structures should also be expected to make a contribution. However, these hypothesis do not exclude the possibility that proteins not forming part of lipoprotein complexes will occasionally exhibit UV fluorescence in the cell. For example, the bright fluorescence of muscles is determined by the proteins of the actomyosin complex [Shtrankfel'd (1963), (1964*a*–*d*)]. It is not known whether the fluorescence of bacteriophages and bacteria is due to lipoproteins,

*The authors embedded the mitochondria in a film of polyvinyl alcohol, so that the objects were not native, and were perhaps dead. No data were given concerning any check of the biological activity of the mitochondria.

though the bacteria contain lipoprotein complexes in mitochondrial structures [Biryuzova and Meisel' (1964)].

The question of the physicochemical nature of the intense fluorescence of lipoprotein complexes remains open to this day. It may be supposed that in such complexes the protein molecule is kept more rigid because the protein and the lipid are intimately associated [Green (1964)]. Such increased rigidity of the molecule should increase the fluorescence yield. It may also be postulated that the yield is enhanced by the electrostatic and hydrophobic bonds linking the proteins and the lipids. For example, it is known that oriented polar groups are observed in phospholipids [Green and Fleischer (1962)], so that on interaction between proteins and phospholipids the polarity of the region near the protein may change, leading to a change of the fluorescence (see Section 1.3). The interaction between proteins and lipids can also result in a redistribution of the changes in the protein molecule [Green and Fleischer (1962)], which might be reflected in the fluorescence (see Section 1.3).

The Nature of Changes in the Intensity of Cell UV Fluorescence

If practically the whole of the cell UV fluorescence is due to proteins, then changes in the cell emission in the course of life processes or under the action of various factors may be associated with changes in the protein concentration, in the screening of proteins by UV-absorbing (but nonfluorescent) substances, and in the conformations of protein macromolecules [Barenboim *et al.* (1961)].

For example, the increase in the fluorescence of blood leucocytes during phagocytosis can well be associated with increased synthesis of proteolytic enzymes [I. E. Brumberg and E. M. Brumberg (1964*a*)], i.e., with an increased protein concentration. The change in cell fluorescence produced by X rays or UV radiation may, in particular, be compared with the change in the screening of proteins in the radiational degradation of DNA; this is discussed in the work of Barenboim *et al.* (1961), (1963*a*).

However, it seems to us that the main reason for changes in cell emission are conformational changes of the cell proteins. This is supported by the following points:

1. The cell fluorescence is due to proteins.
2. The universal response of a cell to an external action is a reversible protein reaction [Nasonov and Aleksandrov (1940); Nasonov (1959)] or, in the language of modern molecular biochemistry, conformational changes of the proteins.
3. Even relatively slight conformational changes are reflected in the fluorescence quantum yield of the proteins (see Sections 1.2, 1.3, and 1.4).

We shall now turn to some specific cases in which the cell UV fluorescence is due to mitochondrial proteins complexed with lipids.

It is known that very diverse factors induce one and the same type of reaction in mitochondria – swelling and shrinkage – accompanied by structural changes of the mitochondrial membranes [see Mashanskii (1965)].

The fact that changes in the UV fluorescence of mitochondria are connected with structural changes was noted by Shudel' (1965*a*) using direct microscopic observations, including electron-microscopic observations, and later by Konev and Lyskova (1965*c*) in whose experiments changes in the respiration of fractions of isolated mitochondria following the addition of succinic acid led to a modification both of UV fluorescence and of light scattering associated with the size of the mitochondria. Succinic acid is known to cause swelling of the mitochondria [Hanter (1962)]. Apart from proteins bound to respiratory coenzymes, mitochondria contain other proteins which might exhibit fluorescence, e.g., structural proteins reacting hydrophobically with phospholipid micelles to form a protein–lipid network or the actomyosin-like protein (AM protein). It is thought that apoenzymes and structural proteins, and also AM protein, undergo conformational changes, contraction of mitochondrial membranes and swelling of the mitochondria [Lehninger (1962*a*), (1964); Green (1964); Neifakh (1964)].

Until more detailed experimental results become available we are entitled to postulate that the conformational changes of the above-mentioned three types of proteins connected with the swelling of mitochondria determine the changes in the mitochondrial emission in response to various factors.* The contribution made by each of these types of proteins to mitochondrial fluorescence has not as yet been determined. The AM proteins could play an appreciable part in the emission since the actomyosin complex exhibits intense fluorescence [Burshtein and Suslova (1964); Shtrankfel'd (1964*a–c*)], possibly owing to the high tryptophan content [for the amino acid composition of AM proteins see, for example, Poglazov (1965)]. In this context it may be noted that, according to Konev and Bobrovich (1965) the polarization spectra of the fluorescence of mitochondria correspond to the spectra of tryptophan-rich proteins.

In his analysis of the action of respiration poisons on fluorescence, Shudel' (1965*a*) discusses as a possible cause of the changes in cell fluorescence the conformational changes in respiratory enzymes. However, since respiration and the action of nearly all such poisons are accompanied by mechanical and chemical alterations in the mitochondrial membranes [Lehninger (1962*a*);

*Konev and Lyskova (1965) found that the addition of urea to mitochondrial suspensions differing in respiration rate reduced the differences between the fluorescence intensities of these suspensions. This was regarded as direct proof of the fact that fluorescence intensity is influenced by the tertiary and quaternary structures of the protein, which are affected by urea. However, it is also possible that the above authors observed not a direct action of urea on the proteins responsible for the fluorescence but an indirect one, since in the concentrations used urea can cause swelling of the mitochondria.

Hanter (1962)], the effect of various cell respiration inhibitors on mitochondrial fluorescence should not necessarily be connected solely with the effect of these poisons on enzyme proteins of the electron-transfer chain.

Application of UV Fluorescence Microscopy to the Investigation of Cells and Tissues

It may be said that it was pure luck that cell fluorescence is in most cases due to the mitochondria, one of the most labile cell organelles and highly sensitive to changes in the condition of the cell. Thus, changes in various mitochondrial characteristics can be used to assess differences in the physiological state of the cell, to determine the early stages of cell alteration, etc. Very diverse factors, such as hormonal preparations, heavy metal ions, fasting, drugs, poisons, or action of sublethal temperature, evoke one and the same type of reaction from the mitochondria – swelling or shrinkage [see the review by Mashanskii (1965)].

The UV fluorescence of cells is thus sensitive to mechanical influence [E. M. Brumberg *et al.* (1961)]; changes in the oxygen concentration (reported by the present author); heating [I. E. Brumberg and Dunaeva (1964)] or cooling [I. E. Brumberg (1964)]; action of urethane, chloral hydrate, and methanol [I. E. Brumberg and Dunaeva (1964)]; action of metabolic poisons [E. M. Brumberg and I. E. Brumberg (1964); Shudel' *et al.* (1964)]; X rays [Barenboim *et al.* (1961)], and ultraviolet radiation [E. M. Brumberg and Barskii (1961); E. M. Brumberg *et al.* (1960*b*)].

Since radiation poisoning of a cell begins at the atomic-molecular level, and the quantum yield (or intensity) of the fluorescence is also a molecular characteristic, it may well be expected that changes in the UV fluorescence of cells would show up early stages of radiation poisoning.

Such changes were in fact observed [Barenboim *et al.* (1961); Fig. 1.23] on leucocytes irradiated with doses of only 42 rad. The dependence of cell fluorescence on the state of the mitochondria also promotes early detection of radiation poisoning, since radiation quickly affects mitochondrial structure and biochemistry [Bacq and Alexander (1961); Manteifel' and Meisel' (1962)]. In fact, Barenboim *et al.* (1963*a*) observed that in isolated mitochondrial fractions changes in fluorescence preceded biochemical changes.

Similar changes in cell emission have been observed in the course of various processes occurring in cells and organisms, for example, mitosis [E. M. Brumberg *et al.* (1961*c*); Chernogryadskaya *et al.* (1964*c*); see also Fig. 1.24], changes in the level of differentiation [Bresler *et al.* (1964*a*); Bresler and Vorob'ev (1965)], embryogenesis and postnatal development [Pil'shchik and Nikoleava (1963); Pil'shchik *et al.* (1964); Pil'shchik and Kudryavtseva (1965)], regeneration [Shudel' (1964*a*), (1965*a*); Chernogryadskaya *et al.* (1965)], passage of an excitation pulse along a nerve fiber [Ungar and Romano (1962)],

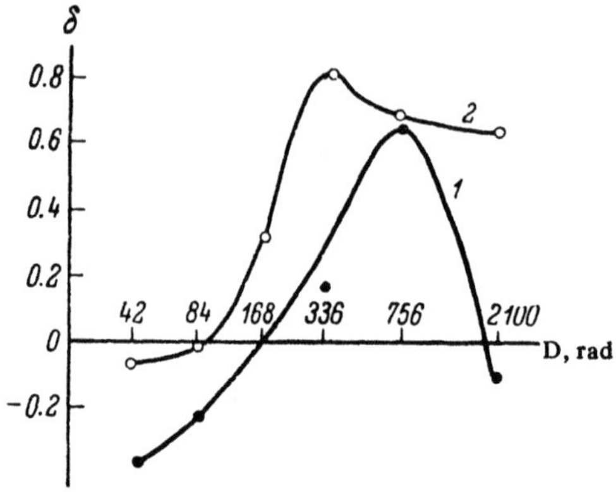

Fig. 1.23. Influence of irradiation on fluorescence of leucocytes [Barenboim *et al.* (1962)]. *D* is the dose of irradiation, plotted on log scale. δ represents the relative effect 1 hr after irradiation. (1) lymphocytes; (2) segmented nuclei of neutrophils.

"survival" of muscles *in vitro* [V. E. Barskii *et al.* (1962); Shtrankfel'd (1963), (1964*a*–*d*)], primary action of a carcinogen on the cell [Bresler *et al.* (1964*a*); Bresler *et al.* (1965*a*, 1965*b*)], and so on.

The fluorescence intensity changes in tumor and leukosis cells [E. M. Brumberg and Barskii (1961); I. Ya. Barskii and I. E. Brumberg (1962); Meisel' *et al.* (1962); Rozanova *et al.* (1962, 1963); I. E. Brumberg and E. M. Brumberg (1964*b*); Shudel', (1964*b*); Bresler (1965)].

In most of these cases the luminescence data were analyzed only qualitatively; the changes in fluorescence indicated that something was changing in the cell. This qualitative character of the analysis of the results is due to historical reasons: the development of UV fluorescence microscopy led to work on the determination of structural and biochemical fluorescence carriers in the cell.

In a number of recent works [see, for example, Bresler *et al.* (1964*a*); Shudel', (1965*a*)] study of the UV fluorescence is used as one of the tools of cytological research, side by side with cytospectrophotometry, measurement of enzyme activity, study of the ultrastructure of mitochondria, light scattering, etc. This combination of various approaches makes it possible to obtain quantitative results and utilize to the full the advantages of UV fluorescence microscopy. Moreover, a connection is today being traced between the numerous data on the dependence between luminescence and protein conformation on the one hand, and data on cell UV luminescence on the other hand. It may be hoped that after precise identification of the proteins responsible for cell fluorescence it will be possible to turn to the connection between UV fluorescence and the state of the protein molecule. Luminescence analysis will then make it possible to obtain information on conformational changes of protein molecules in the living cell.

A special case of the application of UV fluorescence methods to cells is the investigation of the exogenous accumulation of aromatic amino acids by cells. In the authors' laboratory [see Barenboim *et al.* (1967*d*)], the exogenous accumulation of L-tryptophan by the yeast *Candida utilis* which synthesizes tryptophan when one of its precursors (anthranilic acid, indole, indolyl-3-pyruvic and

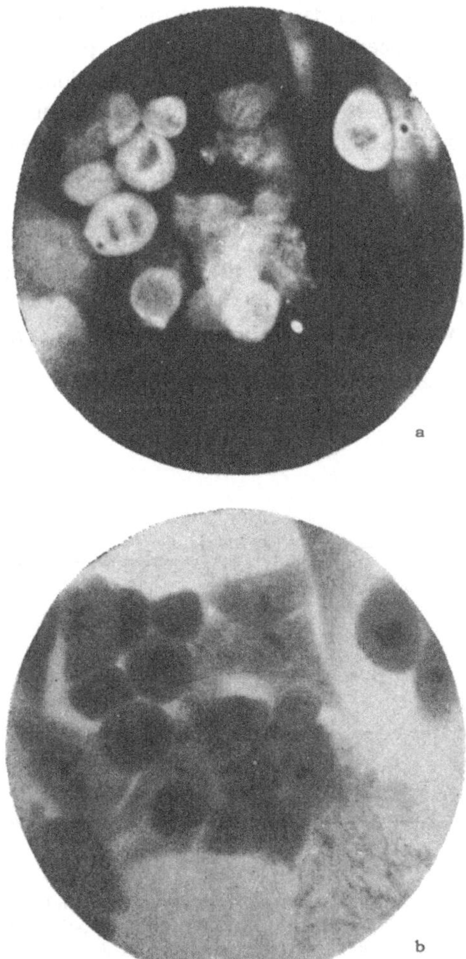

Fig. 1.24. Cultured tissue cells undergoing mitosis. (a) Photographed by the light of their own UV fluorescence (320–400 mμ); and (b) by transmitted light (250–280 mμ).

indolyl-3-lactic acids) is introduced into the medium and accumulates it in the medium in substantial amounts, was studied in detail by a luminescence method. In all these cases, both the precursors and the product of synthesis fluoresce. This process was investigated by the authors in the greatest detail using anthranilic acid as the precursor.

Figure 1.25 shows the general course of the process, recorded by luminescence and absorption methods (initial anthranilic concentration 10^{-4} *M*). The threshold concentrations of anthranilic acid at which no accumulation of tryptophan was observed, either on account of the extremely small amount in the medium or on account of destruction of the acid, were also determined in the work. These concentrations were equal to 10^{-5} *M* and 5×10^{-2} *M* respectively. For initial anthranilic acid concentrations lying between the threshold values, the process of disappearance of anthranilic acid and accumulation of tryptophan was characterized by the numerical values of the rates of these process, the time at which anthranilic acid began to disappear and tryptophan began to be detected, and the efficiency of the conversion of anthranilic acid to tryptophan. These parameters depend on the initial anthranilic acid concentration: as it increases, the processes associated with exogenous accumulation are suppressed.

In the same work a similar picture of tryptophan synthesis from indole

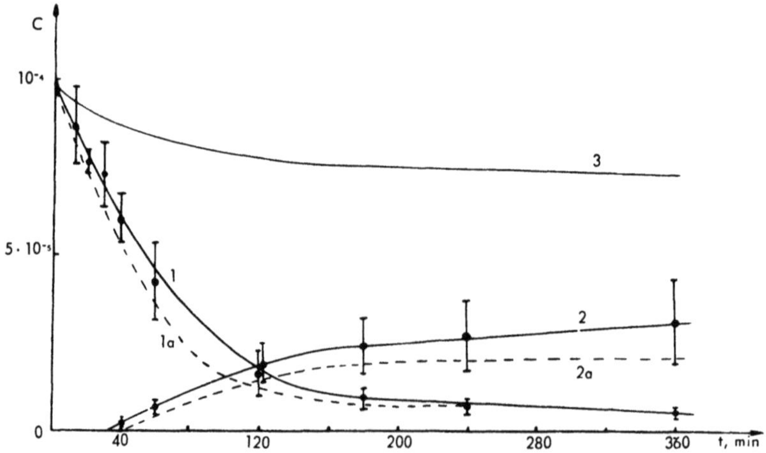

Fig. 1.25. Kinetics of processes associated with exogenous accumulation of tryptophan by a culture of yeast *Candida utilis* in the presence of anthranilic acid (10^{-4} *M*). Along x-axis: time from addition of anthranilic acid (for curves 1 and 2) or tryptophan (for curve 3); along y-axis: concentration in moles. (1) Decrease in anthranilic acid; (2) accumulation of tryptophan; (3) change in content in medium of tryptophan (10^{-4} *M*) added to cells in absence of anthranilic acid. Curves la and 2a (analogous to curves 1 and 2) were obtained in one of the experiments by a spectrophotometric method.

was observed; its quantitative characteristics naturally differed from the case of anthranilic acid. Fluorescence investigations also indicated that compounds that are suitable in principle for tryptophan synthesis, but are not characteristic metabolites for the given strain (indolyl-3-acetic acid, indolyl-3-propionic acid, indolyl-3-butyric acid), do not stimulate exogenous accumulation of tryptophan. They may not arise in the cell on account of the absence of suitable permeases. Also interesting are data indicating that exogenous synthesis is stimulated only by precursors in the chain of synthesis from anthranilic acid on to tryptophan: shikimic acid, for example, which precedes this portion of the chain, does not lead to the accumulation of tryptophan in the medium; the authors attribute this to the allosteric inhibition of anthranilate synthetase by excess tryptophan.

On the whole, such a system ("precursors–yeast–tryptophan") may be considered as a convenient model for studying the most varied aspects of the synthesis of low-MW substances, both in the normal state and under the most varied influences on cells, using UV fluorescence.

Chapter 2

LUMINESCENCE OF PURINES, PYRIMIDINES, AND NUCLEIC ACIDS

Nucleic acids are high molecular weight compounds made up of alternating carbohydrate and phosphate groups

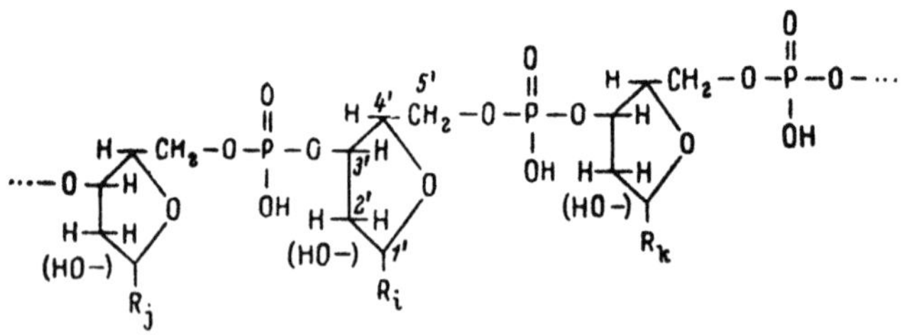

Two types of nucleic acids may be distinguished: desoxyribonucleic (DNA) and ribonucleic (RNA), the corresponding carbohydrate residues being desoxyribose and ribose. The difference between the latter two is that ribose contains a hydroxyl group in position 2′ in place of hydrogen. The main chains of the nucleic acids are otherwise identical.

The side radicals R, whose sequence determines the primary structure of the nucleic acid, are simple derivatives of purine and pyrimidine:

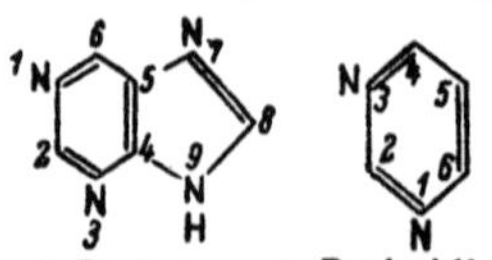

a Purine b Pyrimidine

The purines and pyrimidines are joined to the sugar residues via N_9 and N_1 atoms respectively.

DNA contains two purine bases (adenine and guanine) and two pyrimidines (cytosine and thymine). In RNA the thymine is replaced by uracil.

The formulas for these bases are given below, in keto and amino forms:*

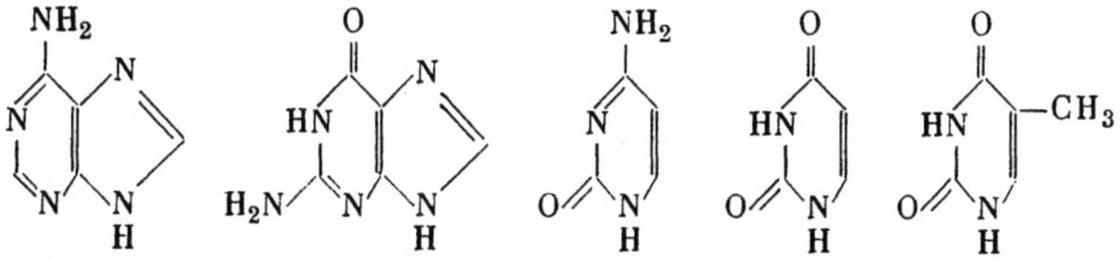

a Adenine b Guanine c Cytosine d Thymine e Uracil

Other purine and pyrimidine derivatives are encountered much more rarely in nucleic acids.

Compounds consisting of a purine or pyrimidine base and a carbohydrate residue are called nucleosides. Addition of a phosphate group to the carbohydrate residues of nucleosides results in the formation of nucleotides.

Purine and pyrimidine bases are not confined to nucleic acids but appear in a whole range of low molecular weight biologically important compounds. The biologically active adenine compounds play a special part. An example is adenosine triphosphate (ATP), which acts as an energy carrier in living systems.

Native DNA consists of a double helix† made up of two polynucleotide chains. The whole structure resembles a ladder twisted around its long axis. The nitrogenous bases play the part of rungs, and the alternating ether-bonded phosphoric acid and desoxyribose residues the part of the handrails. The two main chains of DNA are complementary, in the sense that the adenines and guanines of one chain correspond to thymines and cytosines of the other. The base sequence in one chain (the primary structure) therefore determines unambiguously the primary structure of the second chain.

The polynucleotide chains of DNA are bound by hydrogen bonds between the nitrogenous bases. Formation of pairs A–T and G–C corresponds to the maximum number of intermolecular hydrogen bonds between the chains, and moreover satisfies the steric condition – equality of the DNA "rungs" in length – necessary for a regular structure.

The nature of the forces stabilizing the DNA helix is still not quite certain.

*Nitrogenous bases may undergo tautomeric keto–enol and amine–imine transformations:

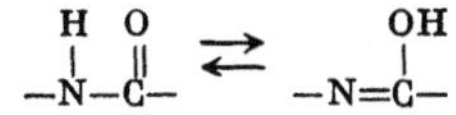

a keto form | b enol form

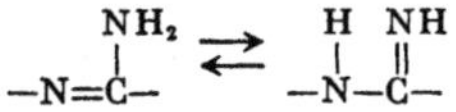

c amino form d imino form

The equilibrium concentration of a given form depends on temperature and pH of the medium. Under normal conditions (room temperature, neutral pH) the bases are predominantly in the keto and amino forms. The tautomerism of purines and pyrimidines is discussed in greater detail by Pullman and Pullman (1963).

†There is at least one exception to this rule – the single-helix DNA of bacteriophage φx174.

The DNA structure is evidently determined not only by the hydrogen bonds but also by van der Waals and hydrophobic interaction forces between the nitrogenous bases.

In spite of its rigid helical structure the DNA molecule does not behave as an ideal rigid rod in solution. It may flex where the spiral structure is disturbed, and consequently it assumes the configuration of a very loosely wound random coil.

Although the chemical structures of DNA and RNA are very similar, there are important differences between their macromolecular structures. In contrast to DNA, the RNA molecule consists of a single polynucleotide chain. Its secondary structure is the alternation of double-helical and single-chain regions. The helical regions are stabilized by intramolecular hydrogen bonds between the purine and pyrimidine bases (pairs A–U and G–C). The RNA structure is very sensitive to external conditions and cannot be regarded as a fixed native structure: in an external medium, of low ionic strength and at room temperature, the double-helical sections orient themselves perpendicularly to the long axis of the macromolecule and pile up; this results in a rod-like conformation. In a medium having higher ionic strength the RNA molecules become less asymmetric and resemble compact coils. At very low ionic strength, or at elevated temperature, the helical sections straighten out and the molecule passes into a loose random coil. Detailed descriptions of the macromolecular structure and the functional significance of nucleic acids will be found in monographs by Bresler (1963) and Volkenshtein (1965), in Tikhonenko's (1965) review devoted to DNA, and in Spirin's (1963) book on RNA.

2.1. THE LUMINESCENCE OF PURINES AND PYRIMIDINES

The luminescence of purine and pyrimidine bases has long attracted the interest of investigators. An example is the study carried out by Stimson and Renter as early as 1941, in which the emission detected in the visible region and apparently due to impurities present in the preparations was observed visually.

Such studies are not only important for an understanding of the luminescence properties of nucleic acids, but are of intrinsic interest in view of the important part played by compounds containing purine and pyrimidine derivatives in various life processes ranging from respiration to muscular contraction.

The first investigations of the intrinsic luminescence of the bases contained in nucleic acids with spectroscopic instruments and photomultipliers date back only thirteen years.

Shore and Pardee (1956) investigated the nucleosides and nucleotides of the bases of nucleic acids in neutral aqueous solutions, as well as thymidine, adenosine-5-phosphate, cytosine, adenylic acid, and adenosine in 0.1*N* HCl and 0.1*N* NaOH. Visible or UV fluorescence was never observed, probably owing to the low sensitivity of the instruments. Weber (1957) also failed to detect fluorescence from adenine and adenylic acid solutions, though his apparatus was more sensitive.

UV fluorescence of adenine, guanine, and their nucleotides and nucleosides was first recorded by Duggan *et al.* (1957). Analysis of subsequent studies shows that the luminescence of purines and pyrimidines appears on excitation (1) below 300 mμ, (2) around 300 mμ, and (3) at 350–360 mμ. The properties of the observed emission depend on the region of excitation. It should be noted that after excitation in any of these regions the luminescence was observed only under certain conditions. This may be due to the specific features of the emission on excitation in different regions of the spectrum or to the impurity of the experimental material.

We shall now consider in turn the characteristics of the luminescence of purines and pyrimidines on excitation in various regions.

Luminescence on Excitation Below 300 mμ

The chromophores responsible for the optical properties of purines and pyrimidines are the purine and pyrimidine rings. These compounds and their derivatives absorb in the ultraviolet. The absorption coefficients of these compounds are several times higher than that for tryptophan, the most strongly absorbing amino acid. Much as been published on the absorption spectra of the nitrogenous bases and nucleic acids. Details will be found in the reviews by Beaven *et al.* (1955) and McLaren and Shugar (1964).

The absorption spectra of a large number of nucleosides and nucleotides are given in an atlas by Venkstern and Baev (1965). Figure 2.1 shows the absorption spectra of five nitrogenous bases for the purpose of general orientation.

Since the biologically important purine and pyrimidine derivatives are formed by the addition of substituents directly into the rings, their absorption (and in particular their luminescence) is largely determined by the nature of such substituents.

The ability of these compounds to undergo protolytic reactions, tautomeric transformations, and various interactions with the formation of hydrogen bonds and charge-transfer complexes, etc., makes the fluorescence properties strongly dependent on the experimental conditions.

Fluorescence of Aqueous Solutions of Purines and Pyrimidines at Room Temperature. This problem has been studied most often. The results obtained by Börresen (1963*a*) and some other authors are shown in Table 2.1. Absolute values of the fluorescence quantum yields of a large number of purines and pyrimidines were determined in Börresen's work. In a few cases these results may be compared with the results of other authors [Udenfriend (1962); Udenfriend and Zaltzman (1962); Walaas, (1963)], and such comparison shows the presence of some disagreement. The wavelengths of the fluorescence maxima given in Table 2.1 are also slightly different from those obtained by Udenfriend

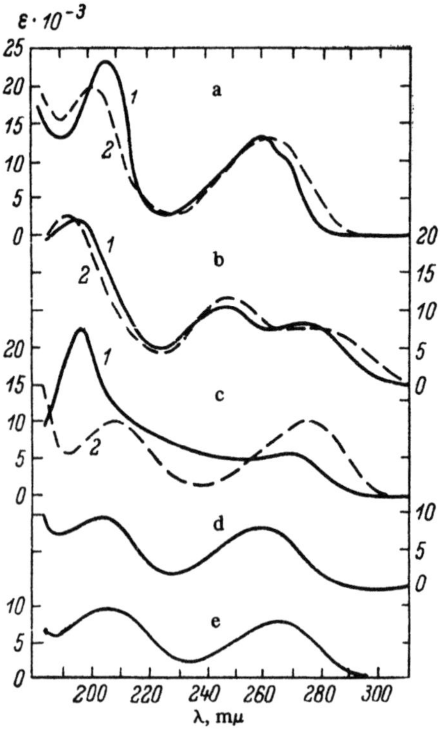

Fig. 2.1. Absorption spectra of the bases of nucleic acids [Voct *et al.* (1963)]. (a) Adenine: (1) at pH 7.0; (2) at pH 2.5; (b) guanine: (1) at pH 6.2; (2) at pH 1.6; (c) cytosine: (1) at pH 8.8; (2) at pH 2.5; (d) uracil, at pH 7.0; and (e) thymine, at pH 7.0.

(1962), Udenfriend and Zaltzman (1962), and Walaas (1963). These discrepancies may be due to the difficulty of recording the weak emissions of the compounds in question, to insufficient purity of the preparations, or to insufficient accuracy of corrections for the spectral sensitivity of the recording part of the apparatus.

However, inspection of Table 2.1 allows us to draw certain conclusions about the fluorescence properties of purine and pyrimidine derivatives. We shall now consider in greater detail the fluorescence properties of the biologically important derivatives of these bases.

Adenine and its derivatives exhibit appreciable fluorescence only in acid media and do not fluoresce at neutral or alkaline pH. According to Udenfriend and Zaltzman (1962) the fluorescence maximum of adenine in water at pH 1 lies at 380 mμ, and the quantum yield is 0.002. Transition to a neutral medium is accompanied by a general fall of the yield (Fig. 2.2). Bőrresen (1963*a*) established that for adenine the pK obtained by the fluorescence method (pK_{fl} – see Chapter 1) on transition from an acid to a neutral medium corresponds closely to the pK obtained by other methods, and concluded that the pH dependence of the fluorescence intensity of adenine is due to a protolytic reaction. It is still

Table 2.1. Characteristics of the Flourescence of Purine and Pyrimidine Derivatives in Aqueous Solutions at Room Temperature

Substance	pK		Fluorescence quantum yield	Fluorescence maximum (mμ)
	Fluorescence titration	Other methods		
Adenine	4.2	4.15; 9.8	0.003 0.002 (pH 1)* 0.0015 (pH 1)**	380
Adenosine	3.7	3.63	0.003 0.005 (pH 1)**	390
Adenosine monophosphate (AMP)	3.95	3.74	0.004	400
Adenosine diphosphate (ADP)	4.3	3.95	0.006	390
Adenosine triphosphate (ATP)	4.6	4.0	0.011	400
2-Amino-4-methylpyrimidine	4.0 12.5	4.15 –	0.062	350
2,6-Diaminopurine	5.05 10.7	5.09 10.77	0.022 (pH 8)	340 (pH 2) 350 (pH 9)
Guanine	3.4 9.3 12.7	3.3 9.2 12.3	0.005 (pH 2) 0.024 (pH 10.7) 0.03 (pH 11)*	355 (pH 2) 340 (pH 11) –
Guanosine	2.2	2.2	0.007	390
Guanosine monophosphate (GMP)	2.5 9.4 12.7	2.4 9.4 –	0.008 (pH 1) 0.008 (pH 10.6)	380 (pH 1) 340 (pH 10.8)
Guanosine diphosphate (GDP)	3.1 –	2.9 9.6	0.018	385
Guanosine triphosphate (GTP)	3.4 –	3.3 9.3	0.010	380
Purine	3.2 9.2	2.39 8.93	0.008 (pH 2.46) 0.0018 (pH 8) 0.045 (pH 10.6)	400 (pH 2) 380 (pH 5.2) 370 (pH 11)
Pyrimidine	2.0	1.3	0.007	360
Thymine			0.015 (pH 11)*	380 (pH 11)*

This table is an abbreviation of the table given by Börresen (1963*a*), supplemented by some data taken from Udenfriend and Zaltzman (1962)(*) and Walaas (1963)(**).

Unless the pH is indicated in parentheses it corresponds to the value at which the maximum fluorescence intensity was observed. For some of these compounds the corresponding pH may be determined from Fig. 2.2.

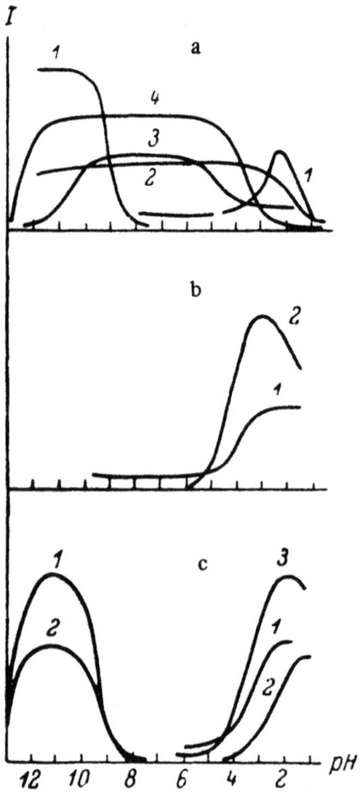

Fig. 2.2. The pH dependence of the relative fluorescence intensity (*I*) for some purine and pyrimidine solutions [Börresen (1963*a*)]. (a): (1) Purine; (2) pyrimidine; (3) 2,6-diaminopurine; (4) 2-amino-4-methylpyrimidine; (b): (1) adenine; (2) ATP; (c): (1) guanine; (2) GMP; (3) GTP.

uncertain which of the groups in adenine the proton attaches to under acid conditions. Pullman and Pullman's (1963) calculations, X-ray analysis of the adenine cation [Cochran (1951); Broomhead *et al.* (1962)], influence of the protonation of adenine on its IR spectra [Tsuboi *et al.* (1962)], and a number of the other data all indicate that the proton adds to the nitrogen atom in position 1. This was also the conclusion reached by Börresen (1963*a*) on the basis of his hypothesis about the mechanism of the action of proteolytic reactions on the fluorescence of purine and pyrimidine bases.

According to other authors, in an acid medium it is the amino group of adenine that undergoes protonation ($NH_2 \rightleftharpoons NH_3^+$). Thus, Walaas (1963) supported this view by reporting that fluorescence yield of the solution of inosine monophosphate, differing from AMP by the replacement of the NH_2 by an OH group on C_6 of the purine ring, is constant from pH 1 to pH 10. However, this cannot be taken as reliable proof because the replacement of NH_2 by OH can change extensively the whole electronic structure of the molecules and thus also its fluorescence properties.

To elucidate the possible changes in the fluorescence properties of adenine on its inclusion into DNA molecule it was of considerable interest to study the

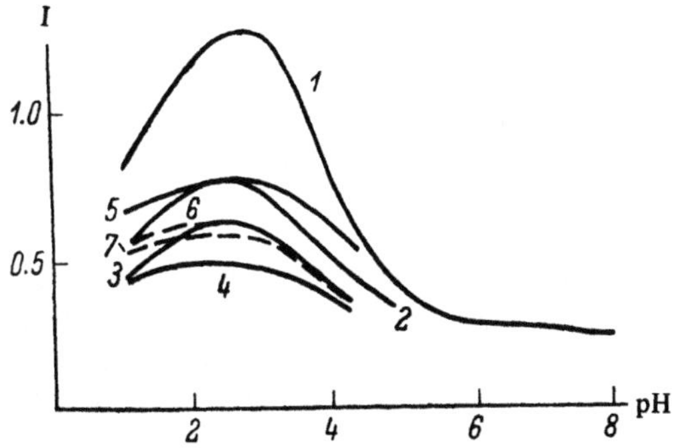

Fig. 2.3. The pH dependence of the relative fluorescence intensity (*I*) of adenine and its derivatives [Watanabe *et al.* (1963)]. (1) ATP; (2) ADP; (3) AMP; (4) adenosine; (5) adenine in 1.0 *M* $(CH_3)_4NCl$; (6,7) ATP and ADP in 1.0 *M* $(CH_3)_4NCl$ and 0.5 *M* $MgCl_2$.

properties of simple adenine derivatives – its nucleosides and nucleotides. This was done primarily by Watanabe *et al.* (1963).

By analyzing the pH dependence of the fluorescence intensity of adenine, adenosine, AMP, ADP, and ATP, and also by studying the effect of metal ions on this dependence (Fig. 2.3) the above authors came to interesting conclusions about the configuration of adenine nucleotides in solution. In their view the negatively charged phosphate groups interact directly with the purine ring, and this is why the nucleotides fluoresce more strongly in comparison with adenine and adenosine. Such ring configuration of nucleotides is disturbed in the presence of metal ions (Ca^{2+} and Mg^{2+}), which compete against the purine ring in the bonding with phosphate groups.

According to Walaas (1963) at pH 1 the fluorescence yield of adenine and its derivatives decreases in the order adenosine ⟩ ATP ⟩ ADP ⟩ AMP ⟩ adenine, which is in disagreement with Watanabe *et al.* (1963).

Guanine and its nucleosides and nucleotides are the most intensely fluorescing purine derivatives. Like adenine, guanine does not fluoresce in neutral solutions, but does so increasingly at higher and lower pH (Fig. 2.2). The greatest quantum yield for guanine is at pH 11. Under these conditions $q = 0.03$ and the fluorescence maximum lies at 350 mμ [Udenfriend and Zaltzman (1962)]. In acid media, q reaches its maximum (0.005) at pH 2, and the fluorescence maximum is observed at 335 mμ [Börresen (1963*a*)].

The addition of a carbohydrate ring and phosphate groups to guanine leads to a sharp change in the fluorescence properties. Udenfriend and Zaltzman (1962) report that guanine nucleosides and nucleotides practically do not fluoresce at alkaline pH, but this disagrees with Börresen (1963*a*) according to whom the fluorescence quantum yield of AMP at pH 10.6 is 0.008. Thus, here too there is a certain discrepancy between the results of different authors.

It is interesting that methylation of the amino group in guanine results in compounds exhibiting very strong fluorescence. The emission intensity of a solution of dimethylaminoguanine, present in large amounts in transfer-RNA, at pH 11 exceeds the emission intensity of neutral tryptophan solutions [Udenfriend and Zaltzman (1962)].

Of the pyrimidine bases of nucleic acids, only thymine exhibits appreciable fluorescence. Its quantum yield in alkaline media is 0.0015. Even the nucleoside and nucleotide form of thymine do not fluoresce in solution [Udenfriend and Zaltzman (1962)].

Fluorescence of Purine and Pyrimidine Solutions at Low Temperatures. The fluorescence properties of purines and pyrimidines in frozen solutions differ markedly from their fluorescence properties at room temperature. Longworth (1962) reported on the fluorescence of purine, adenine, and guanine in polar frozen 1:1 mixtures of propane-1,2-diol and water at 77°K and pH 7. Under these conditions the fluorescence maximum of the purine spectrum lies at 297 mμ, that of adenine at 229 and 309 mμ, and that of guanine at 323 mμ; their quantum yields are 0.02, 0.06, and 0.09, respectively. The fluorescence spectrum of adenine under these conditions has a vibrational structure.

Particular mention should be made of the fact that at low temperature the maximum fluorescence yield is observed at neutral pH [Rahn *et al.* (1964); Fig. 2.4]. This has been confirmed by Börresen (1963*b*), who showed that the fluorescence of ATP in a 1:3 water–methanol mixture at pH 3.5 practically disappears when the temperature is lowered. Börresen explained this by postulating that the emission of the fluorescence of this compound must be preceded by transition of the molecule from state $n\pi^*$ to state $\pi\pi^*$, which requires thermal activation. Other explanations may also be put forward. In particular, it could be proposed that the temperature change affects the equilibrium of the various tautomeric forms of the molecule. The influence of the solvent composition on the fluorescence properties of the bases must also be accounted for. Study of the temperature dependence of the fluorescence intensity of bases at various pH may provide information about the energy levels of the molecules.

Fluorescence of Purines and Pyrimidines in the Solid Aggregate State. No one has yet studied the fluorescence of such purine and pyrimidine preparations at room temperature, but some results for 90°K and for excitation at

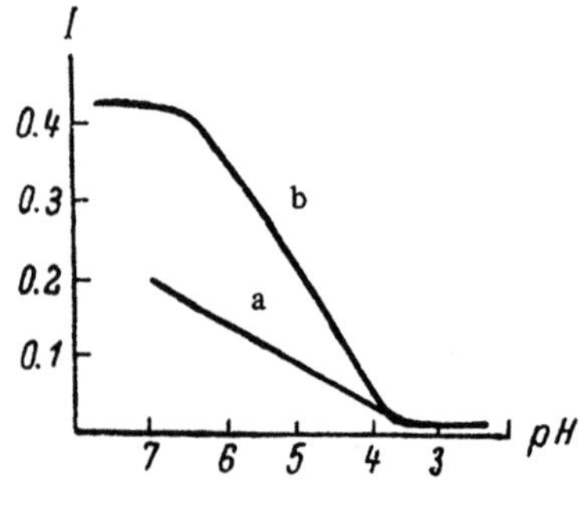

Fig. 2.4. The pH dependence of (a) the fluorescence intensity and (b) the phosphorescence intensity of AMP in frozen solutions (77°K) [Rahn *et al.* (1964)].

260–280 mμ are reported by Agroskin *et al.* (1960). The most interesting fact is that both purine and pyrimidine bases give an emission under these conditions (Fig. 2.5).

Considering the work of Agroskin *et al.* (1960) on the purine and pyridine bases in the solid aggregate state on deep cooling, we may ascribe the appearance of the emission to the low temperature or to the influence of the solid aggregate state. The latter possibility follows, for example, from I.Ya. Barskii's (1959) report that the fluorescence intensity of guanine absorbed on paper is much higher than in solution. Thus there are some grounds for supposing that weak fluorescence may already be detected in solid specimens of purines and pyrimidines, and that this emission is enhanced by cooling.

Phosphorescence of Frozen Purine and Pyrimidine Solutions. Phosphorescence of purine, adenine, and guanine in frozen solutions (at 77°K) at pH 7 was observed by Longworth (1962). The phosphorescence spectra lie at 360–500 mμ, with peaks at 400, 388, and 410 mμ; the quantum yields of the above bases are 0.038, 0.045, and 0.036, respectively. The phosphorescence spectra of adenosine and guanosine are shown in Fig. 2.6. The phosphorescence spectra of adenine and its derivatives are strongly structured. The intensity is greatest in neutral media and falls off on transition to acid media [Rahn *et al.* (1964)].

The decay times of the phosphorescence of guanine and its derivatives, and of adenine and its derivatives, are 1.2 and 2.3 sec respectively [Bersohn and Isenberg (1963, 1964)].

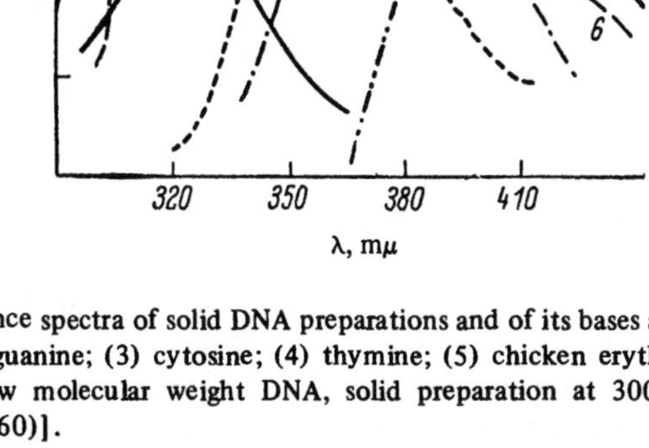

Fig. 2.5. Fluorescence spectra of solid DNA preparations and of its bases at low temperature. (1) Adenine; (2) guanine; (3) cytosine; (4) thymine; (5) chicken erythocyte DNA; all at 90°K; and (6) low molecular weight DNA, solid preparation at 300°K [(1)–(5) after Agroskin *et al.* (1960)].

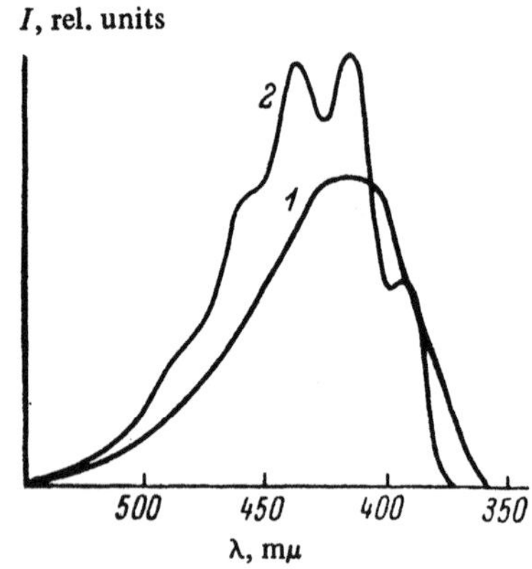

Fig. 2.6. Phosphorescence spectra of (1) guanosine and (2) adenosine [Bersohn and Isenberg (1963)].

Luminescence on Excitation in the Region of 300 mμ

Only the fluorescence of solid purines and pyrimidines cooled to 90°K has been studied with excitation in the region of 296–313 mμ [Agroskin *et al.* (1960)]. Luminescence was observed in both cases. A highly interesting observation is that on excitation in this region the emission spectra of thymine and uracil were displaced in comparison with the spectra excited at 260–280mμ, from 310 to 395, and from 355 to 430 mμ, respectively. In the case of guanine, adenine and cytosine the positions of the emission maxima remained unchanged (at 370, 355, and 400 mμ); see Table 2.2.

The very fact that luminescence could be excited at around 300 mμ where there is no appreciable absorption by the free bases, deserves special attention. It also still unexplained why the fluorescence spectra of frozen uracil and thymine preparations shift depending on the spectral region of the excitation.

Luminescence on Excitation in the Region of 350 mμ

In addition to the main absorption maximum at 260–285 mμ in the spectra of concentrated aqueous solutions of adenine, adenosine, inosine, thymidine, uracil, cytosine, thymine, and hypoxanthine, Basu and Greist (1963) found an absorption "shoulder" at 350 mμ. The intensity of this "shoulder" increases with increasing concentrations of these solutions and with a rise in their viscosity (the viscosity was increased by addition of glucose to the solutions).

Illumination of concentrated solutions by light with a wavelength of 350 mμ both in the presence and absence of glucose gives rise to visible fluorescence with a maximum at 450 mμ (Fig. 2.7). The intensity (*I*) of this emission is very low, and is slightly higher in the presence of glucose. The fluorescence

Table 2.2. Fluorescence Maxima of the Spectra of Solid Preparations of Purine and Pyrimidine Bases and of Their Derivatives at 90°K [Agroskin *et al.* (1960)]

Substance	Excitation range 260–280 mμ	296–313 mμ
Adenine	355	355
Adenosine	375	375
Adenosine-5-monophosphate	360	350
Guanine	370	370
Guanosine	325	355
Guanosine-5-monophosphate	370	360
Cytosine	400	400
Cytidine-5-monophosphate	330	340
Uracil	355	430
Uridine-5-monophosphate	320	430
Uridine-5-triphosphate	330	340
Thymine	310	395
Thymidine-5-monophosphate	335	375

yield increases with increasing concentration of the solution. In strongly diluted solutions, light of the same wavelength did not excite any fluorescence. The emission band observed in concentrated solutions is most intense in neutral media, falling off considerably at pH below 4 and above 8. The intensity of the absorption and emission bands of adenosine, inosine, and uracil is the highest; these bands are ascribed by Basu and Greist to the formation of molecular change-transfer complexes in concentrated solutions.

2.2. THE LUMINESCENCE OF NUCLEIC ACIDS

This has been much less studied than the luminescence of proteins, mainly because the emission intensities are low and it is difficult to obtain luminescence-pure preparations of nucleic acids. As in the case of the nitrogenous bases, the emission of nucleic acids appears on excitation in three spectral ranges – below 300, around 300, and at 360–370 mμ.

Luminescence on Excitation Below 300 mμ

Solutions. The first attempt to study the fluorescence of nucleic acids in aqueous solutions was made by Shore and Pardee (1956). These authors studied neutral aqueous solutions of RNA from tobacco mosaic virus and from yeasts, and of DNA from salmon sperm, but failed to detect any emission in the visible and UV regions. These results were confirmed by Agroskin *et al.* (1961): on excitation in the usual absorption region nucleic acids do not fluoresce in neutral aqueous solutions. At the same time Agroskin *et al.* (1961) observed emission with a maximum around 358 mμ from 0.1 *N* HCl (pH 1.1) solutions of nucleic acids of various origins. Study of the excitation and fluorescence spectra of nucleic acids and their comparison with the corresponding spectra of purine and pyrimidine bases allowed the above authors to conclude that the fluorescence of DNA and RNA is under these conditions due mainly to guanine; the contribution of adenine is many times smaller, and pyrimidines do not contribute at

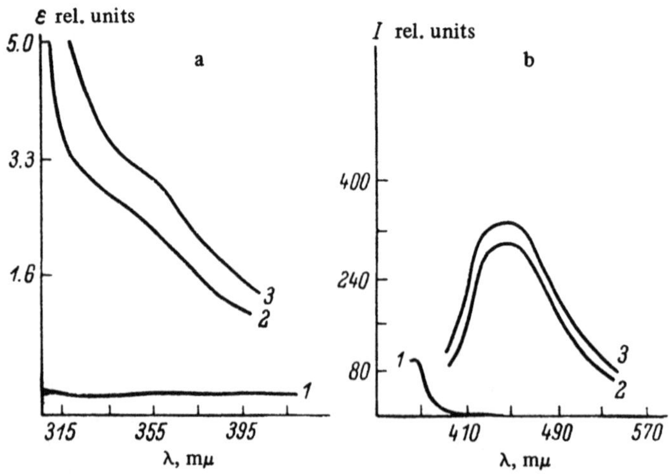

Fig. 2.7. (a) Absorption and (b) fluorescence spectra of adenosine solutions of various concentrations [Basu and Greist (1963)]. (1) 0.0003 *M* aqueous solution; (2) 0.03 *M* aqueous solution; (3) the same solution with an addition of sucrose. λ_{exc} = 350 mμ. The pH of the solutions was 7.0

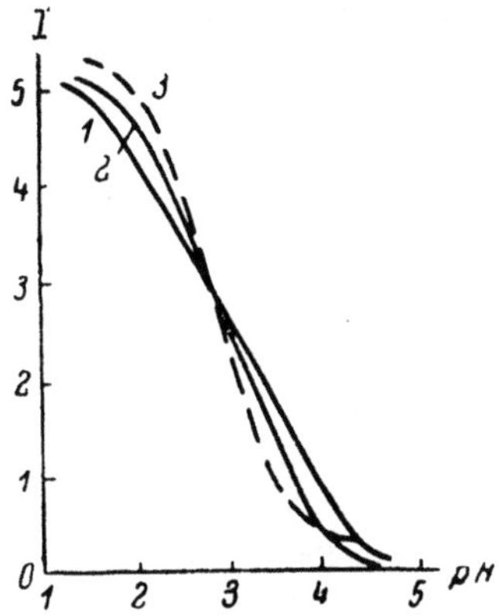

Fig. 2.8 Effect of the pH of the medium on the fluorescence intensity (*I*) of aqueous solutions of DNA and guanine. (1) Low molecular weight DNA; (2) high molecular weight DNA; (3) guanine.

all to the emission. In good agreement with this conclusion, purine-free nucleic acid does not fluoresce, while pyrimidine-free acid does and gives a spectrum similar to that of DNA.

The depolymerization of DNA in solution at pH 1.1 [Rice and Doty (1957)] reduces the value of the results reported by Agroskin and co-workers because the latter authors were in fact observing the fluorescence from depolymerization products. For this reason Barenboim *et al.* (1963*b*) measured the pH dependence of the fluorescence intensity of DNA to record the emission of high-polymeric DNA (Fig. 2.8).

The figure shows that the curves for DNA and guanine are very similar. The fact that DNA fluoresces only at low pH was ascribed [Agroskin *et al.* (1961)] to fluorescence of salts of the purine bases in the acid medium, in particular salts of guanine, the base which determines the fluorescence of DNA. However, according to our data guanine hydrochloride does not fluoresce in neutral aqueous solution.

It seems more likely that the pH dependence of DNA emission is caused by ionization of the guanine residues. Fig. 2.8 shows that the fluorescence of guanine in DNA is half-quenched at pH 3, which is approximately the pK value for guanine (Table 2.1). It is not known to which group in guanine this pK refers, but Pullman and Pullman (1963) suggest that it is connected with the transfer of the proton at N_7 in guanine-forming part of nucleic acids. In any event, the theory that the pH dependence of the DNA fluorescence is connected with a protolytic reaction is in good agreement with the ideas developed in the previous section, and in particular with Borresen's (1963*a*) results.

In contrast to the case of proteins, the structure of DNA does not influence the fluorescence quantum yield of the chromophore. This conclusion may be reached on the basis of the fact that the transition of DNA from the helical to the coil conformation [which takes place at pH$<$ 3.5 according to

Sukhorukov *et al.* (1963)] has practically no effect on the pH dependence of the emission intensity (Fig. 2.8).

The absence of fluorescence from DNA solutions on excitation in the region of usual absorption under physiological conditions and the independence of the fluorescence of nucleic acids on their structure weakens considerably the interest in the study of the fluorescence of these compounds.

It is worth noting that one may utilize the UV fluorescence from DNA and RNA excited at $\lambda < 300\,m\mu$ for a quantitative determination of their contents of guanine and its derivatives. In particular, the fluorescence method has been used for a quantitative analysis of RNA hydrolysates [Udenfriend and Zaltzman (1962)]. It is known that there are no appreciable amounts of free bases in alkaline hydrolysates of RNA, and consequently such preparations fluoresced most strongly in acid solutions. The authors consider that the weak fluorescence of the alkaline hydrolysates of RNA at pH 11 is due to the splitting off of small amounts of free guanine under these conditions. However, on the basis of Börresen's (1963*a*) results on the emission from GMP in alkaline media (Table 2.1) the fluorescence of alkaline hydrolysates of RNA can be explained by the presence of this compound.

In contrast, acid hydrolysates of RNA fluoresce most strongly in alkaline media [Udenfriend and Zaltzman (1962)] owing to the presence of large amounts of free guanine. This conclusion is also confirmed by the similarity between the fluorescence spectra of such acid hydrolysates and guanine. Udenfriend and Zaltzman (1962) developed a fluorescence method for estimating guanine quantitatively in RNA hydrolysates, and the results obtained by this method are in good agreement with chromatographic results, though the luminescence method is both simpler and much quicker.

The same authors (1963) also proposed a method for analyzing small amounts of dimethylaminoguanine in the presence of a large quantity of guanine (e.g., in the hydrolysates of transfer-RNA). The method is based on investigation of the fluorescence of the mixture of these two purines after treating the mixture with nitrous acid. Dimethylaminoguanine is unchanged by HNO_2, while guanine becomes transformed into xanthine which does not fluoresce under these conditions. The method gives results that agree well with those obtained by other techniques (for example, chromatography).

Fluorescence of Frozen Solutions. Rahn *et al.* (1964) investigated the fluorescence of DNA and synthetic polynucleotides in a 1:1 mixture of ethylene glycol and water at 77°K. Under these conditions only the purine polynucleotides fluoresce; the spectral maxima of poly-A and poly-G were observed at 335 and 360 mμ respectively. The fluorescence intensity of DNA was about a tenth of the total intensity due to its component purine nucleotides. Rahn and coworkers found that the fluorescence of the polynucleotide chain depends fundamentally on the latter's structure. Thus, denaturation of DNA enhances the

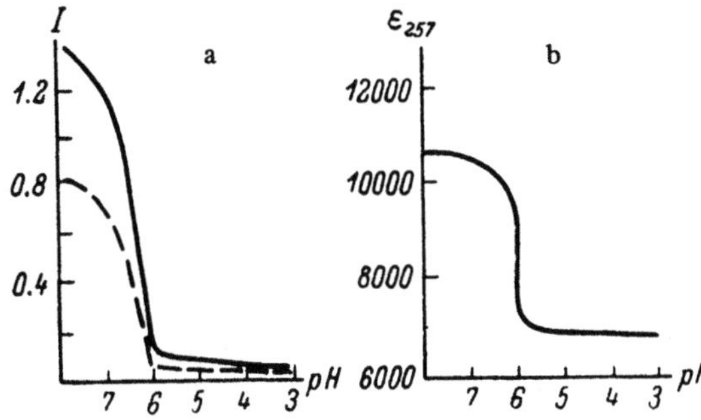

Fig. 2.9 pH dependence of the relative intensity of fluorescence and phosphorescence (*I*) and of the absorption coefficient (ε) of polyadenylic acid at 77°K [Rahn *et al.* (1964)]. (a) Fluorescence and phosphorescence; (b) absorption coefficient.

emission by a factor of about three. A structure dependence of fluorescence was also observed in experiments with poly-A. Figure 2.9 shows the relative fluorescence intensity of this polymer as a function of pH. A sharp fall in intensity occurs around pH 6. Rahn *et al.* (1964) explained this by a co-operative transition of poly-A from the random coil conformation (pH 6) to the helical form (pH<6). The occurrence of such a transition at pH 6 is known from a number of works [see, for example, Steiner and Beers (1958)], and was also confirmed in the above publication by Rahn and co-workers: at pH~6 there is a sharp decrease in the absorption coefficient (hypochromic effect) at 257 mμ (Fig. 2.9).

Rahn and co-workers further report that the fluorescence of the nucleotides of adenine and poly-A is quenched on the formation of triple helices, which appear on an interaction of a poly-A molecule with two molecules of poly-U, or a molecule of poly-A with two molecules of polyinosine. Since the spiral conformation of poly-A molecules is possible only when the NH_2 groups of adenine are ionized, the quenching of the fluorescence may be due to the influence of these groups on the chromophore.

Fluorescence of Nucleic Acid Preparations in the Solid Aggregate State at Low Temperatures. Agroskin *et al.* (1961) found that solid preparations of DNA and RNA cooled to 90°K fluoresced on excitation at 260–280 mμ, giving a wide fluorescence band that covers the entire spectral region of the emission due to the nitrogenous bases. The "shoulders" on the curves of the fluorescence spectra of nucleic acids characterize the contributions of individual bases to the emission (Fig. 2.5). The difference between the spectra of DNA and RNA is due to the difference between the contributions of thymine and uracil, bases that are specific to these nucleic acids.

Luminescence on Excitation in the Region of 290–313 mμ

In 1960 Agroskin and co-workers detected fluorescence from solid preparations of nucleic acids cooled to the temperature of liquid oxygen and excited

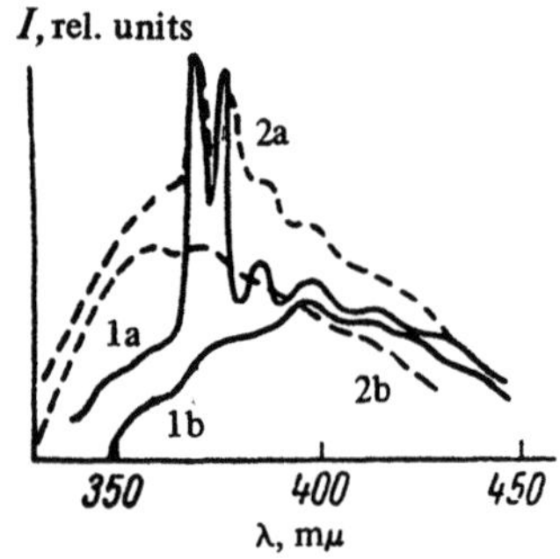

Fig. 2.10. Fluorescence spectra of RNA and DNA preparations [Agroskin *et al.* (1960)]. (1) RNA from tobacco mosaic virus; (2) DNA from chicken erythrocytes; (a) before and (b) after UV irradiation for 30 min. $\lambda_{exc.}$ = 296–313 mμ.

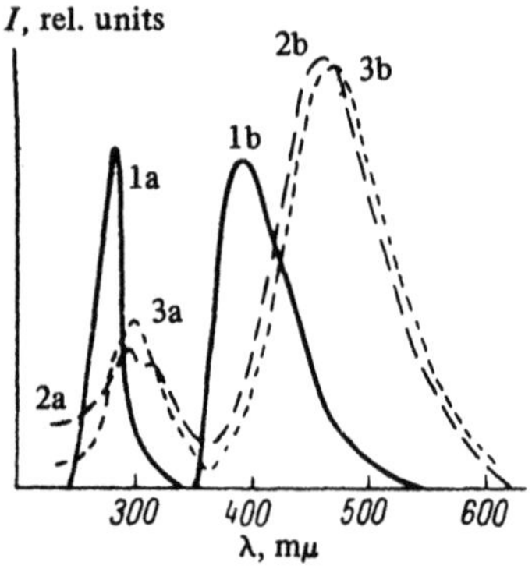

Fig. 2.11. Excitation and emission spectra of (1) histone, (2) nucleoprotein, and (3) DNA [Douzou *et al.* (1961)] (a) excitation (b) emission.

with light at 290–313 mμ (maximum at around 300 mμ). This is the same excitation region as that of the nitrogenous bases (Table 2.2). One of the most interesting results of this work was the discovery of a fine structure in the fluorescence spectra of RNA from tobacco mosaic virus and DNA from chicken erythrocytes (Fig. 2.10). Some time later Douzou *et al.* (1961) also observed fluorescence and phosphorescence of frozen DNA solutions (Fig. 2.11). An important characteristic of this emission was the dependence of the excitation and luminescence spectra and of the decay time on the structure of the molecule and on interaction with proteins, especially histones (Figs. 2.12–2.14). Finally, in 1961 Agroskin and his associates described the fluorescence of DNA and RNA solutions in 0.1*M* phosphate buffer at pH 7.2, with an emission maximum at 300 mμ. However, no more detailed studies have been reported. The luminescence excited in this region can clearly be of considerable interest for the study of the structural transformations of nucleic acids, but no suggestions have as yet been put forward about its nature, except that it is connected with "structural centers."

Fluorescence on Excitation with Light of Wavelength Around 350 mμ

Basu and Lox (1962) recorded a "shoulder" in the absorption spectrum of neutral solutions of nucleic acids at 350 mμ and emission in the visible region (with a maximum at 440 mμ) on excitation with light in the region of 350 mμ (Fig. 2.15). The emission died out after prolonged exposure of the solutions to ultrasound. Subsequent work [Basu and Greist (1963); Basu and Lox (1963)] led the authors to the conclusion that the long-wave "shoulder" in the absorption spectrum and the new fluorescence band in the visible part of the spectrum are due to the formation of complexes between the bases of nucleic acids. The character of these complexes is not clear, though the authors consider that a variety of charge-transfer complexes is involved. Special additional experiments are evidently necessary to explain the nature of this emission; these

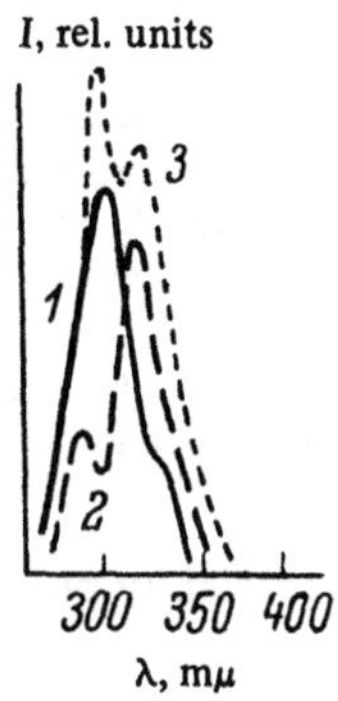

Fig. 2.12. Changes in the excitation spectra of DNA after heating or irradiation [Douzou *et al.* (1961)]. (1) Native DNA; (2) DNA after 4 hours at 40°C; (3) DNA irradiated for 2 hours with ultraviolet light at 77°K.

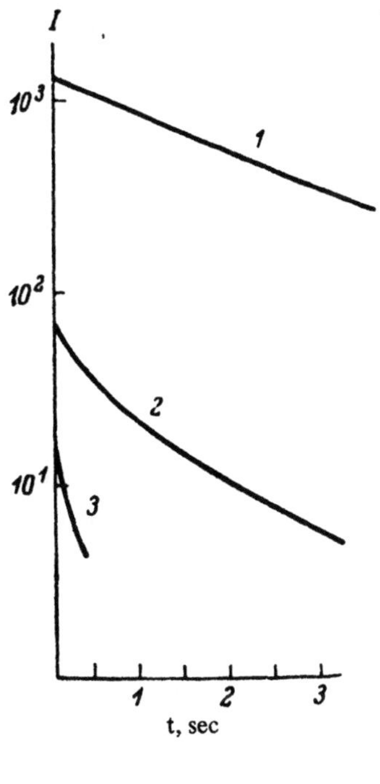

Fig. 2.13. Decay of the luminescence of polynucleotides and DNA [Douzou *et al.* (1961)]. (1) Polyadenylic acid, $\lambda_{exc.}$ = 300 mμ, $\lambda_{lum.}$ = 430 mμ; (2) polyuridylic acid, $\lambda_{exc.}$ = 310 mμ, $\lambda_{lum.}$ = 460 mμ; (3) DNA, $\lambda_{exc.}$ = 300 mμ, $\lambda_{lum.}$ = 475 mμ.

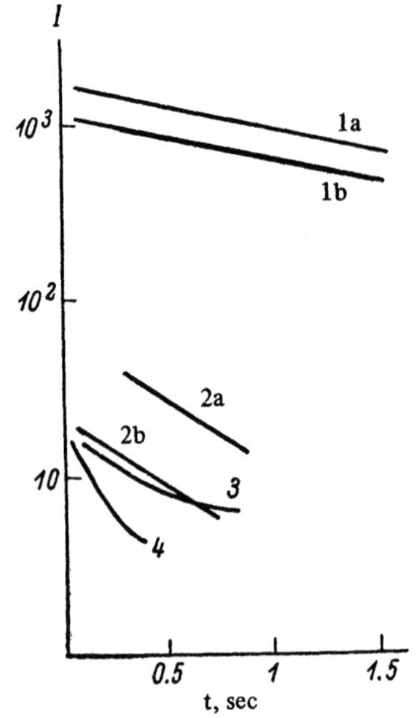

Fig. 2.14. Decay of the luminescence of nucleohistones in DNA [Douzou *et al.* (1961)]. (1) Histone, $\lambda_{exc.}$ = 280 mμ, $\lambda_{lum.}$ = 395 mμ; (2) nucleohistone, $\lambda_{exc.}$ = 300 mμ, $\lambda_{lum.}$ = 445 mμ; (3) nucleohistone, $\lambda_{exc.}$ = 335 mμ, $\lambda_{lum.}$ = 475 mμ; (4) DNA, $\lambda_{exc.}$ = 300 mμ; $\lambda_{lum.}$ = 475 mμ. (a) and (b) represent different experiments.

should be done when we are confident that we are dealing with sufficiently pure preparations of the nucleic acids.

Gachokovskii (1960) discovered a wide, continuous fluorescence band emitted by solid specimens of nucleic acids on excitation in the region of 360 mμ. It is possible that this may have the same nature as the emission of nucleic acid solutions on excitation near 350 mμ [Basu and Lox (1962)].

Luminescence and the Purity of Nucleic Acid Preparations

Influence of Possible Admixtures. While in the luminescence analysis of proteins extensive use is made of well-developed modern methods of purification, such as chromatography, electrophoresis on starch, and preparative ultracentrifugation, in the case of nucleic acids it is difficult to point out any work in which such necessary purification had been done. It is interesting to note [Udenfriend and Zaltzman (1962)] that a number of prêparations regarded as "pure" from the point of view of normal biochemical tests contained sufficient impurities to change appreciably the excitation and luminescence spectra.

In addition, many of the impurities present in nucleic acids are difficult to remove: it will suffice to recall the series of studies carried out by Blumenfeld (1959), concerned with the discovery and investigation of the anomalous magnetic properties of DNA, which proved to be due to bonding of ions to certain segments of DNA, though the latter had been subjected to careful purification. This example is also important because the ions (atoms) of iron are known to be active luminescence quenchers. Similar remarks may be made about the effect of Ca^{2+} and Mg^{2+} ions on the luminescence of nitrogenous bases [Watanabe *et al.* (1963)]. If impurities exert an appreciable influence on the luminescence of nucleic acids, it follows that the presence of impurities can be assessed on the basis of luminescence. In our view the appearance of a peak at 280 mμ in the excitation spectra of nucleic acids may serve as a sensitive indication of the presence of protein admixtures.

"Individuality" of Preparations. Practically all the works on the luminescence of nucleic acids were concerned with the emission of "gross" nonfractioned preparations of DNA and RNA. To appreciate more clearly just what this means, let us for the moment turn back to the proteins. Studying the luminescence of egg albumin, for example, we assume that the emission comes from molecules identical in their amino acid composition and sequence, and having the same spatial structure. On the other hand, when Agroskin *et al.* (1961) measured the fluorescence of RNA isolated from yeast, he was in fact dealing with the combined fluorescence of a mixture of messenger, ribosomal, and transfer-RNA molecules, differing in compositon, chain length, and structure. Such results will naturally reflect only the laws common to a fairly large class of molecules; all effects of molecular individuality will be lost. Conversely, the observation of the fine structure in the fluorescence spectra of RNA from tobacco mosaic virus was favored by the fact that each particle of TMV contains only one RNA molecule, so that the investigated solution contained molecules of one and the same type (if we assume that the isolation of the RNA was sufficiently mild to avoid appreciable fragmentation of the molecule).

Apart from such loss of individuality of the luminescence characteristics in work with a set of nucleic acids of various types, in some cases a small proportion of luminescing molecules may go undetected. It would thus be interesting to study the fluorescence of individual fractions of transfer-RNA. The increased content of the methylated derivatives of nitrogenous bases in transfer-RNA [see the review by Baev (1965)] may lead to the result that the fluo-

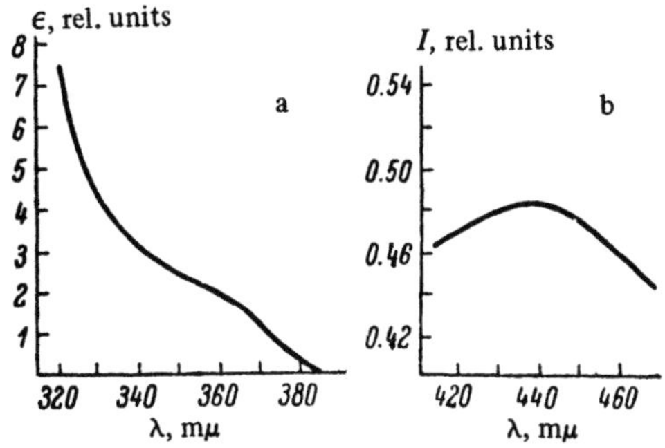

Fig. 2.15. Absorption and fluorescence spectra of a concentrated solution of DNA [Basu and Lox (1962)]. (a) Absorption and (b) fluorescence spectra. Solution concentration 0.8% $\lambda_{exc.}$ = 350 mμ. (*I*) is the fluorescence intensity and ε the absorption coefficient.

rescence quantum yield of these macromolecules excited by UV ($\lambda < 300$ mμ) will be much higher than that in nonfractionated preparations of RNA.

It is thus clear that the luminescence properties of nucleic acids have received much less attention than the luminescence properties of proteins. This is largely due to experimental difficulties: the recording of weak light intensities, the necessity for careful purification of the preparations, and strong light scattering by solutions of nucleic acids.

As regards the nitrogenous bases, luminescence can evidently provide the same information as in the case of the aromatic amino acids: in the first place information about their electronic structure and properties of the excited states.

The absorption spectra of various purine and pyrimidine bases differ little from one another, while the luminescence properties are highly individual. This makes it possible to use luminescence for qualitative and quantitative analysis of nucleic acids.

Work published in recent years demonstrates that there is a connection between the luminescence and the structure of nucleic acids.

Studies of the luminescence of nucleoproteins, a field that is still practically unexplored, will undoubtedly prove of considerable interest.

Chapter 3

RADIOLUMINESCENCE OF BIOPOLYMERS AND THEIR COMPONENTS

A wide variety of inorganic and organic substances luminesce under the effect of X rays or radiation from radioactive sources; these substances include proteins, nucleic acids, and their components. Reports on the radioluminescence of DNA will be found in monographs by Tarusov (1955) and Dubinin (1961). Sapezhinskii and Emanuel' (1960) recorded an after-emission of proteins and nucleic acids under the action of γ rays. Zommermayer *et al.* (1961) observed luminescence from aqueous solutions of aromatic amino acids exposed to soft (7 keV) X rays. An analogous luminescence from crystalline and dissolved tryptophan under more severe conditions of excitation (60 keV) was investigated by Barenboim and Domanskii (1963). Augenstein *et al.* (1964*b*) and Carter *et al.* (1965) studied the X-ray luminescence of powders of aromatic amino acids and proteins.

The observed emission is due to a whole set of processes occurring by various mechanisms arising as a result of an interaction between the X or γ radiation and the substance. Therefore we must, even if only briefly, consider the basic features of these processes and assess their contribution to the emission before coming to the main laws and characteristics of radioluminescence.

3.1. PRIMARY AND SECONDARY EXCITATION PROCESSES

The propagation of X and γ radiation in any medium is accompanied by interaction of the radiation with the substance in question. Over a dozen different types of elementary reactions between γ rays and matter are known; their probabilities depend on the photon energy and on some physical characteristics of the substance [Fano (1953*a*)].

In most radiobiological experiments the radiation has an energy of 0.01–

2 meV; the main interaction processes in this energy range are the photoelectric effect and Compton scattering [Hine and Brownell (1956)], giving rise to the first generation of secondary electrons. These electrons can then undergo elastic and inelastic scattering. The latter effect can result in molecular ionization and excitation, and also in bremsstrahlung (though in the energy range under consideration the losses due to bremsstrahlung may be neglected for substances having a density close to unity: [Nelipa (1960)]).

Ionization by first-generation secondary electrons gives rise to the next generation of electrons and some of these can in turn induce ionization and excitation. On irradiation with 100 keV X rays the number of ionizations due to secondary electrons is 600 times as large as the number of ionizations caused directly by the X rays [Morrisson (1952)]. If the primary particles are electrons, the secondary electrons are responsible for about 50% of the total number of ionizations [Lea (1959)].

The energies of the secondary electrons range from thermal energies up to the energies of the primary particles. The secondary electrons are arbitrarily divided into fast, capable of causing ionization and excitation (δ electrons, $E >$ 100 eV), and slow, incapable of ionizing molecules [Fano (1953*a*)]. Most of the secondary electrons are slow: only 5% have energy greater than 100 eV and 1% energy greater than 500 eV [Charlesby (1960)].

Thus the interaction between radiation and matter gives rise to a large number of low-energy electrons which, as will be seen later, play the main part in molecular excitation.

Direct Excitation

Electrons regardless of origin can induce excitation of molecules of the medium on inelastic collisions. In contrast to photons, electrons having any energy exceeding the excitation energy are capable of exciting a molecule.

The true mechanism of the electron excitation of complex molecules in the condensed phase is unknown, but in a number of works the excitation process is discussed using the laws derived for gases, although such simple extrapolation may be insufficient [Fano (1961); Lipsky (1964)]. In this case the probability of optically permitted molecular or atomic transitions into an excited state n under the action of an electron is determined by the effective cross section Q_n which depends on the properties of the molecule or atom and on the kinetic energy T of the electron [Hart and Platzman (1961)]

$$Q_n = \frac{4\pi a_0 R^2}{T}\left(\frac{f_n}{E_n}\right) \ln (C_n T), \tag{3.1}$$

where a_0 is the Bohr radius of hydrogen, R is the Rydberg constant, C_n is a constant depending on the excitation level, and f_n is the oscillator strength of the electronic level with energy E_n.

Equation (3.1) indicates that only optically permitted transitions can take place. The probability of such transitions first increases with increasing kinetic energy of the electron, reaches a maximum, and falls off monotonically. Calculation using Eq. (3.1) shows that the excitation is conducted mainly by low-energy electrons.

The probability of the excitation of an electronic level is determined by the oscillator strength (more precisely, by f_n/E_n). In aromatic amino acids the area under the absorption curve (and thus the oscillator strength) at 120–240 mμ is much larger than that at 240–300 mμ, indicating that in this case electrons excite mainly the level corresponding to the absorption band at 120–240 mμ.

The oscillator strength of the peptide bond in proteins in the region of 120–240 mμ is smaller than in tryptophan, but the considerably greater number of peptide bonds in a protein in comparison with the number of aromatic amino acid residues results in the fact that an appreciable part of the energy from the ionizing radiation is used up to excite peptide linkages. If it is taken into account that there is no migration of energy from the peptide bonds to the aromatic amino acid residues [Vladimirov and Konev (1959)] and that the peptide bonds themselves do not luminesce, it becomes understandable why the yield of the radioluminescence of proteins is much smaller than that of amino acids. Thus, according to Augenstein *et al.* (1964*b*), the energy needed to obtain the emission of one photon by the tryptophan residues of trypsin is ~ 60 times the energy required for the emission of one photon by free tryptophan.

Equation (3.1) does not reflect the fact that an important characteristic of direct electron excitation of molecules is the possibility of exciting optically forbidden transitions. For example, Kuppermann and Raff (1964) observed excitation by low-energy electrons of two forbidden transitions in ethylene, one of which was a singlet–triplet ($S \to T$) transition.

One of the possible mechanisms of increasing the probability of exciting forbidden $S \to T$ transitions is the exchange of a molecular electron for an incident electron [Pshezhetskii (1962)]. The level multiplicity may be changed in the course of such an exchange if the spin moments of the incident and the ejected electron are opposed. The probability of excitation of a forbidden transition has a sharp maximum for electrons with energy corresponding to the energy of the transition; at larger kinetic energies of the electron the probability quickly falls off because the electron exchange depends on the time of the interaction [Pshezhetskii (1962)].

The second possible reason for the increased probability of the formation of excited triplet states is that excitation by secondary electrons gives rise to a relatively large number of higher excited singlet states. At the same time, for higher excitation levels the probability of a transition to a triplet may be higher than for lower levels [Robinson and Frosch, cited by Augenstein *et al.* (1964*b*)]. This has been confirmed experimentally by O'Dwyer *et al.* (1962).

Finally, the increased probability of the formation of triplet levels may be due to the fact that the high-energy particles with a short range (e.g., the fast δ electrons) produce in their track a large number of ionized and excited molecules, which are consequently strongly localized and form clusters. It is postulated that the strong interaction of excited molecules in such clusters increases the probability of a change in the level mutiplicity [Fano (1961)].

Thus, on excitation with ionizing radiation we should observe an increased number of triplet excited levels in comparison with their number after photo-excitation. This results in a greater phosphorescence component in the radio-luminescence spectrum as compared with the photoluminescence spectrum [Augenstein *et al.* (1964*b*)].

Secondary Processes

In the primary processes the excited molecules are formed as a result of a direct interaction between an electron and a molecule (in principle, an electron may also excite ionized molecules and other secondary products). The remaining methods of excitation may be called secondary, though chronologically some of them (e.g., ion-electron recombinations) may occur earlier than the excitation by electrons of the last generation still having sufficient energy for this purpose. We shall now consider briefly the indirect means of molecular excitation.

Recombination of Positive Ions with Electrons. This should be favored by the Franck-Rabinovich "cage" effect, preventing the escape of a macromolecular ion, and by the short range of electrons in condensed matter.

The recombination of positive ions with electrons may lead to the formation of both singlet and triplet excited states. W.L. Buck (1960) states that triplet excited states are formed in 75% of the cases and singlet excited states in only 12%.

According to a calculation carried out by Rozman and Kilin (1959) the majority of ions in the condensed phase recombine in 10^{-12} sec, which makes it practically impossible to differentiate between the emission due to the recombination of positive ions and electrons and the emission due to direct electron excitation.

Another special feature of a recombination involving a positive ion may be triple collision as a result of which a second particle becomes excited (e.g., the solvent in the liquid phase).

Reactions of Free Radicals. Ionization of molecules leads to the formation of free radicals whose unpaired electrons endow them with high reactivity: radicals recombine readily with themselves, attack neutral molecules, interact with oxygen, and so on. Some of the reactions involving free radicals formed by the action of X and γ radiation on proteins and nucleic acids lead to the formation of excited molecules (see below).

Apart from the recombinations of electrons with ions and free radical reactions there are also other processes that might in principle contribute to the emission (recombination of oppositely charged stable ions, electron capture by neutral molecules), but the part played by these processes in excitation is relatively insignificant [see the reviews by Pshezhetskii (1962) and Barenboim (1963*a*)].

3.2. RADIOLUMINESCENCE AND ITS MAIN CHARACTERISTICS

The emission initiated by ionizing radiations can be divided into three main forms, depending on the mechanism of its appearance.

1. Emission due to direct excitation of molecules of the medium by secondary electrons; i.e., radioluminescence,* which will be considered later in greater detail.

2. Vavilov–Cerenkov radiation, appearing when an electron is moving in a medium with a velocity that exceeds the velocity of light in this medium, and luminescence induced by this radiation. The minimum electron energy causing Vavilov–Cerenkov radiation is determined by the refractive index of the medium, and is about 260 keV for aqueous solutions of proteins.

The Vavilov–Cerenkov radiation and the resulting luminescence may be of appreciable intensity. Chizhikova (1961) notes that the radioluminescence of solutions of organic scintillators may be comparable with the intensity of the Vavilov–Cerenkov radiation in this medium (excitation by γ rays from Co^{60}) only at relatively high concentrations of the solutions (10^{-3} g/ml). The ratio of luminescence intensity excited directly by electrons and that induced by the Vavilov–Cerenkov radiation varies from 0.27 to 6.25 depending on the substance and the solvent.

3. Emission appearing as a result of indirect excitation during secondary processes (see Section 3.1), which may be termed radiochemiluminescence. The greatest attention has been focussed on the emission associated with the reactions of free radicals.

Sapezhinskii *et al.* (1963, 1964a, 1964b, 1965b) showed that the afterglow following X irradiation of the solutions of various proteins (trypsin, ribonuclease, haemoglobin, etc.) is due to recombination of peroxide radicals of the proteins. The analogous effect in DNA solutions is associated with the hydroperoxides of thymine [Emanuel' *et al.* (1965); Zhizhina *et al.* (1965)]. In principle, such an emission could also be proposed for cells. In this context it is interesting to mention the work of Brumberg and Barskii (1960) who observed visible fluorescence after irradiation of dead cells (fixed preparations) with X rays and UV rays. Dry cells did not exhibit this effect, but an emission appeared after immersion of the cells in water. A similar effect of emission appearing on dissolution of previously irradiated substances has been reported for proteins [Sapezhinskii *et al.* (1963)] and for sugar, starch, and glucose subjected to the action of γ radiation [Westermark and Grapendieser (1960)]. The delayed emission of irradiated proteins was associated in

*This term will be used here independently of the ionizing radiation used for the excitation.

Sapezhinskii's work with the formation and recombination of peroxide radicals of the proteins. Brumberg and Barskii (1960) noted that the emission died out after the addition of strong oxygen acceptors.

The emission arising under the action of ionizing radiation is characterized primarily by its spectral distribution, lifetime, and energy yield. Unfortunately, these parameters are often insufficient to distinguish between radioluminescence and emission appearing as a result of secondary processes. For example, radioluminescence and the emission due to recombination of positive ions with electrons (see Section 3.1) come from low excited levels and are thus spectrally indistinguishable. These two types are also indistinguishable by their lifetimes, since the time of ion–electron recombination (10^{-12} sec) is negligible in comparison with the lifetime of a lower excited state (10^{-9} sec).

Radicals of complex organic molecules may have lifetimes far longer than the decay times of fluorescence and even phosphorescence of biopolymers and their components. The radical recombination process itself can also take place after a time greater than the lifetime of the fluorescence. Therefore the contribution of this process to the total emission intensity evoked by ionizing radiation is determined not only by the actual light yield but also by the duration of the measurement: if one records single scintillations the chemiluminescence processes practically do not appear, while when the radioluminescence intensity is measured over a long time their contribution may be very high.

These remarks indicate that, although radioluminescence is defined as a specific physical process, methodologically it is difficult to separate it from the accompanying emissions caused by the specific features of the interaction between the ionizing radiation and the substance.* Therefore the term radioluminescence is often used in the literature to denote a set of emissions having a lifetime of the order of the lifetime of photofluorescence. Radioluminescence is in this case due mainly to true excitation of molecules by electrons and to ion–electron recombinations.

Spectral Characteristics of Radioluminescence

It has been found [Chizhikova (1961); Birks (1964); Birks *et al.* (1964)] that the photo- and radioluminescence spectra of organic scintillators coincide at room temperature. An analogous result was obtained by Barenboim and Domanskii (1963) after excitation of tryptophan powder with soft (60 keV) X rays (Fig. 3.1). However, Augenstein *et al.* (1964*b*), who excited the luminescence of crystalline tryptophan powder with hard (230 keV) X rays, observed a

*Under certain conditions such accompanying emission may also be thermoluminescence induced by X or γ irradiation.

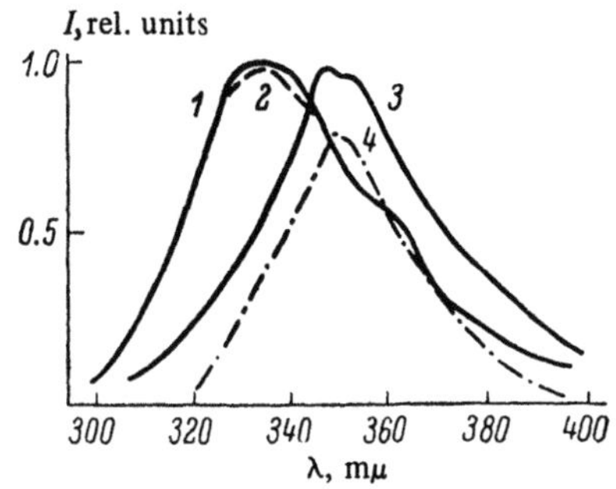

Fig. 3.1. The fluorescence spectra of tryptophan. (1,2) Crystalline tryptophan: (1) excitation by UV, 240–280 mμ and (2) excitation by 60 keV X rays. (3,4) Alcoholic solution of tryptophan, 5×10^{-4} *M*: (3) saturated with nitrogen and (4) saturated with oxygen; in both cases excitation with 60 keV X rays. (*I*) is the fluorescence intensity. The scales on the ordinates are different for spectra 1 and 2 and the same for 3 and 4.

long-wave shift of the maximum* (by about 15 mμ in comparison with photoluminescence) and an additional maximum in the region of 600 mμ.

The coincidence of the photo- and radioluminescence spectra means that the emission is in both cases from the same energy levels. On the other hand, it is possible to envisage excitation of ionized molecules and other secondary products of the interaction of the radiation with the substance; the singlet states of these species are displaced relative to the level of the neutral molecule in the ground state, and this will naturally lead to a shift of the luminescence spectrum.

New impurity-type emission centers may also be formed [Birks (1964)], the part of impurities being played by ions. Finally, some authors [Kallman *et al.* (1955); Belikova (1959)] recorded at room temperature an emission which they believed to be associated with the existence of triplet levels which are formed in much greater numbers under the action of ionizing radiation than on photoexcitation (see Section 3.1).

Augenstein *et al.* (1964*b*) compared the emission spectra of powders of aromatic amino acids and trypsin, appearing upon excitation by 230 keV X rays and by ultraviolet rays at the temperature of liquid nitrogen. Some of their results are shown in Table 3.1.

In tyrosine and trypsin the emission maxima in the second band are close for both kinds of excitation. In the case of tryptophan and phenylalanine the emission maximum after X-ray excitation is shifted towards the red end of the spectrum. In tryptophan this is a two-component maximum and one of the components disappears at room temperature.

Kasha (1964) proposed that the spectral shift observed by Augenstein *et al.* (1964*b*) may also be associated with radiation-chemical changes occurring in

*Since radiation excites luminescence from the surface and X ray from the whole bulk of the powder, the conditions of reflection and light scattering of the luminescence are different in the two forms of excitation. This difference may account for the observed spectral shift, since in Chizhikova's work (1961) different conditions of multiple reflection and light scattering of the luminescence resulted in an apparent displacement of the spectrum.

the substance under the action of X rays, and suggested comparison of the photoluminescence spectra before and after X irradiation.

The spectral characteristics obtained by Augenstein *et al.* (1964*b*) allow comparison of the short-wave and long-wave components of the luminescence spectrum after excitation with X and UV radiation (Table 3.1). The ratios R_1/R_2 and F/P are equal only for tryptophan; for the other compounds $R_1/R_2 < F/P$.

This latter observation is in agreement with the fact that in electron excitation the relative amount of triplet states is much greater than in photoexcitation.

In 1960 Sapezhinskii and Emanuel' recorded an afterglow from frozen solutions of egg albumin, DNA, and RNA exposed to γ radiation. Since the spectra and the decay kinetics of this emission coincided with analogous data for photoexcitation, the authors came to the conclusion that they were dealing with γ phosphorescence. However, the method used did not allow detection of the γ fluorescence of the materials tested, and therefore these results cannot be compared with those shown in Table 3.1.

Lifetime of the Radioluminescence

As was already shown in Section 3.1, radioluminescence is preceded by the formation of the same excited singlet states as during photoexcitation. It might thus seem that this should result in the same lifetimes in the two cases, i.e., $\tau_{photo} = \tau_{radio}$.

Experimental determinations of τ have been conducted for various organic scintillators, but there is still no agreement about the ratio $\tau_{photo}/\tau_{radio}$ and

Table 3.1. Comparison of the Characteristics of Photo- and Radioluminescence of Some Substances at Low Temperature [Augenstein *et al.* (1964*b*)]

Substance	1st band of the spectrum (fluorescence) (λmax)		2nd band of the spectrum (phosphorescence) (λmax)		Ratio of areas under 1st and 2nd bands		Energy absorbed per emitted quantum (eV)
	Photo-excitation (F)	Radio-excitation (R_1)	Photo-excitation (P)	Radio-excitation (R_2)	F/P	R_1/R_2	
Tryptophan	330	355	490	510–560	1.5	1.5	68
Tyrosine	300	320	450	450	1.5	0.013	109
Phenylanaline	295	310	415	500	4.5	0.11	147
Trypsin	330	355	445	455	4.5	0.5	4000

about the character of the decay of radioluminescence [Liebson *et al.* (1950); Chizhikova (1961); Birks (1964)].

Galanin and Chizhikova (1961) have obtained good agreement of τ_{photo} and τ_{radio} for anthracene and naphthalene by comparing the decay of γ scintillation pulses and pulses excited by an electric spark. On this basis Chizhikova (1961) considers that the ratio between the probabilities of radiative and radiationless transitions (which determine the magnitude of the quantum yield) is the same in photo- and radioexcitation. Another group of authors [Kilin and Rozman (1963); Kilin *et al.* (1964)] showed that excitation of liquid and solid solutions with ionizing radiations is accompanied by specific quenching leading to an appreciable reduction of the lifetime (sometimes by a factor of more than 3). The occurrence of additional quenching of the fluorescence from tryptophan and tyrosine, appearing during the action of soft X rays, has been observed by Barenboim and Domanskii (1963). From these studies it follows that the yield of luminescence from a lower excited singlet level is higher during photoexcitation, because of the specific quenching action during excitation by ionizing radiations. Such ideas [Kilin *et al.* (1964)] may be useful towards an understanding of the interaction between ionizing radiation and organic substances, and in particular for establishing the connection between the lifetimes and the yields of photo- and radioluminescence.

Investigations of the low-temperature afterglow of biological objects have been reported by Sapezhinskii and Emanuel' (1960) and by Augenstein *et al.* (1964*b*). The former group recorded the γ phosphorescence of proteins and nucleic acids, and found that τ was identical for each substance with photo- and γ ray excitation. However, the value τ = 6.6 sec obtained by Sapezhinskii and Emanuel' for proteins is more than twice the accepted value [Debye and Edwards (1952)]. Both this and comparison of Sapezhinskii's work with later studies [Sapezhinskii *et al.* (1964*b*, 1965*b*)] suggest that in the case of the afterglow of proteins and nucleic acids the authors might have been dealing also with chemiluminescence arising in the course of irradiation.

Augenstein and other workers discovered that the decay of the afterglow of aromatic amino acids (with the exception of phenylalanine) and trypsin consists of the sum of two exponential decay processes apparently connected with two long-wave components of the spectrum. The phosphorescence of aromatic amino acids after photoexcitation also proved to consist of two exponential components, but both its time of decay and the overall decay were different from the time of decay under the action of X rays.

3.3. YIELD OF RADIOLUMINESCENCE

One of the most important characteristics of radioluminescence is the energy yield or efficiency, i.e., the ratio of energy emitted as radioluminescence

(E_{rl}) to the total absorbed energy (E_{abs}):

$$B = \frac{E_{rl}}{E_{abs}} \cdot \tag{3.2}$$

It is clear from the preceding section that this definition may mean several things, depending on what is understood by radioluminescence. The energy E_{rl} emitted under the action of ionizing radiation is made up of energies emitted from lower singlet and triplet levels ($E_s + E_t$) excited directly by electrons (radiofluorescence and radiophosphorescence) and of the sum of energies E_p emitted as a result of all other processes accompanying true radioluminescence (luminescence due to the recombination of positive ions with electrons, recombination of radicals, etc.). Thus,

$$B = \frac{E_s + E_t + \Sigma E_p}{E_{abs}} = B_s + B_t + \Sigma B_p. \tag{3.3}$$

Each of these component yields depends on the probability on the processes preceding the formation of the corresponding excited state and on the probability of radiation from this level. In the extensive literature on scintillation counters [see the review by Birks (1964)], the radioluminescence yield is usually determined at room temperature after a time commensurable with the scintillation lifetime, under conditions when radio- and photoluminescence spectra coincide and there is practically no afterglow. The main contribution to the radioluminescence yield comes then from direct excitation and recombinations of electrons with ions, leading to emission from a lower excited singlet level. For this case we may write

$$B = \rho q, \tag{3.4}$$

where ρ is the fraction of absorbed energy leading to the formation of a lower singlet state and q is the photoluminescence quantum yield of the substance. Analyzing results obtained from (3.4) we must bear in mind that the probabilities of the radiative transitions from a lower singlet level in radio- or photoexcitation are different because of the additional quenching specific for radiofluorescence (see above), so that the B values calculated from Eq. (3.4) may be somewhat high.

In γ excitation the value of ρ for solutions of the wide range of substances investigated by Chizhikova (1961) varies over a relatively narrow range around $\rho = 3.9\%$. If we assume that for solutions of aromatic amino acids $\rho = 3.9\%$ then for tryptophan, tyrosine and phenylalanine, whose photoluminescence quantum yields are 20, 21, and 3.5% [Teale and Weber (1957)] the yields B will be respectively 0.78, 0.82, and 0.137%. The radioluminescence yields of protein solutions, which like the photoluminescence yields are evidently determined by the aromatic amino acid residues, should also be low. Still lower values may be expected for the radioluminescence yields of nucleic acids, whose photoluminescence yields [Agroskin *et al.* (1961*b*)] do not exceed 1%.

Three methods may be used to determine the energy yield of radioluminescence. In the first the luminescence intensity is measured with the substance exposed to continuous irradiation by a beam of particles or quanta with simultaneous determination of the absorbed energy. In the second – the pulse method – the measured quantity is light energy emitted in single scintillations caused by electrons of known energy. For example, in the case of the γ radiation from Co^{60} the scintillations are due to Compton electrons whose energy may be calculated. Finally, one can determine the relative yield in comparison with a substance whose absolute energy yield had been determined by one of the first two methods. In principle, all three methods may be used for biopolymers and their components, though the scintillation ability of these molecules has never been confirmed, and the experiments may encounter insurmountable difficulties. In practice, only two methods have been used so far: Augenstein *et al.* (1964*b*) determined the absolute radioluminescence yield of deep-cooled powders of aromatic acids and dry proteins,* and Barenboim and Mochalkin (1966) used the relative method.

According to Augenstein *et al.* (1964*b*) the average quantum emitted over the entire range of the emission [ed. note: this includes emission not classified as normal fluorescence] corresponds to 147, 109, and 68 eV of absorbed energy for phenylalanine, tyrosine, and tryptophan, respectively, while the corresponding values for every quantum emitted at the fluorescence peak are 1400, 8350, and 110 eV for the same amino acids. Judging from the fluorescence peaks given in this work (310, 320, and 355 mμ), the energy yield in the peak is 0.32, 0.05, and 3.55% for phenylalanine, tyrosine, and tryptophan. From the fluorescence quantum yields, equal to 0.04, 0.21, and 0.2 [Teale and Weber (1957)], it is easy to calculate that for these amino acids the amounts of energy leading to the excitation of one molecule to the lowest singlet level are 56, 1670, and 22 eV. Since the energy required in the electronic transition from the ground state to the lowest singlet level is only about 4.5 eV for these amino acids, then in fact excitation to this level by high energy radiation in phenylalanine, tyrosine, and tryptophan requres much more initially absorbed energy, since the excited

*Determination of the energy yield by the method selected by Augenstein and co-workers requires measurement of the absorbed dose of the incident radiation and absolute photometric measurement of the intensity of the emitted light. It is known that luminescence yields obtained for powders are not in fact true quantum yields because in powders the unavoidable multiple reflections and scattering leads to reabsorption of the luminescence. This has been demonstrated experimentally, e.g., by Chizhikova (1961) by comparison of data for powders and crystals. Therefore the usual procedure is to use special measurements and calculations for powders [Alentsev and Vinokurov (1951)], whose application in the case of radioluminescence is hindered by the fact that the penetrating radiation excites emission from the whole mass of the powder.

singlet energy represents, respectively, 0.08, 0.003, and 0.2 of the absorbed radiation energy. These results are in qualitative agreement with ours: the radiofluorescence intensity of tyrosine powder was much smaller than that of phenylalanine, while the opposite order was observed for solutions. Barenboim and Mochalkin (1966) compared the energy yield of a tryptophan solution with the yield of a solution of quinine sulfate, for which Chizhikova (1961) determined ρ. All the measurements were conducted with excitation by soft (60 keV) X rays; in this case the solutions have a relatively high radioluminescence intensity and there is no masking effect due to the Vavilov–Cerenkov radiation. The results were then recalculated for energies corresponding to the energy of Co^{60} emission. These experiments show that for tryptophan in solution irradiated with Co^{60} the energy yield should not exceed 0.26%. The differences between the B values obtained for tryptophan by Augenstein *et al.* (1964*b*) and by Barenboim and Mochalkin (1966) may be due to various causes, mainly to the different photoluminescence yields in powders and solutions and to different energies of ionizing radiations corresponding to B in each of these cases.

The radioluminescence energy yield is proportional to the photoluminescence quantum yield [Eq. (3.4)]. In turn, for powdered phenylalanine $q \approx 1$ [Yeargers and Augenstein (1965)] while in solution $q = 0.04$. This sharp rise in q is due to the existence of aggregation which reduces quenching during collision, the absence of luminescence quenching during energy transfer, and other processes that the authors believe to be specific to the solid phase. Therefore we may consider that for trytophan powder too, q will be much higher than for its solution, which can also lead to an increase in B in the former case.

The differences between the values of B obtained for tryptophan by various authors may also be caused by the different energies of X and γ radiations to which these B values correspond. A strong influence of the energy of the primary ionizing radiation on radioluminescence yield has been shown by many investigators* [see the reviews by Kallman and Furst (1958); Chizhikova (1961); Birks (1964)].

3.4. RADIOLUMINESCENCE AND THE ROLE OF EXCITED STATES IN RADIOBIOLOGY

Study of radioluminescence allows certain conclusions to be drawn about the role of excitation to higher molecular states [ed. note: author uses

---

*Analysis of the results reported by Barenboim and Mochalkin (1966) showed that at 0.06 –1.2 MeV the yield will be, in the first approximation, directly proportional to the specific energy loss, and inversely proportional to the magnitude of the energy. Therefore the B values obtained on γ excitation (1.2 MeV) must be multiplied by 1.2/0.23 = 5.2 before being compared with the B obtained on X-ray excitation (0.23 MeV). The yields for tryptophan reduced to the same energy (0.23 MeV) are 3.55 and 1.35%, i.e., they are fairly similar.

"excitation" for transitions to optically allowed molecular states] in the primary processes of the biological action of ionizing radiation.

Is the excitation energy responsible for radiation poisoning, or is the whole effect due to ionization? This has long been discussed in the literature [Lea (1959)] but some authors believe that the role of the processes of excitation in radiation poisoning of the cell is still uncertain [Eidus (1964) Chapter 21].

The participation of excitation energy in radiation poisoning is determined (1) by the energy distribution between ionization and excitation, and (2) by the character of the processes in which a given excited state is reached. Dissipation of the excitation energy in small portions over the large number of degrees of freedom of biopolymers or light emission may lead to energy losses that are harmless to the organism, while predominant formation of triplet forms may favor radiation-chemical processes.

Both these problems (energy distribution and character of the process) have still not been solved experimentally for condensed phases. It is believed that about half of the energy absorbed by the condensed phase is used up on excitation of ions and molecules [see the review by Barenboim *et al.* (1963*a*)], but experimentally this has only been proved for gases [Williams and Essex (1949)], and the extrapolation of the laws for gases to condensed phases is not always valid (see above). For vapors of organic substances Rozman and Kilin (1959) found experimentally that the ratio of excited to ionized molecules N_e/N_i, is equal to 3. According to Platzmann and Frank (1958), in condensed phases $N_e/N_i = 1-2$, and this is also the value obtained by approximate calculations for proteins and nucleic acids [Barenboim (1963*a*)]. If it is considered that about 42.5% of the whole mass of the cell consists of biopolymers with directly bound water [Kuzin (1958)], then 42.5% of the whole absorbed energy of the ionizing radiation will go to such complexes, and half of this will be used for excitation. If we rely on such very rough estimates, we find that the proportion of energy absorbed by the biopolymers is sufficiently large to stimulate us to ask about its subsequent fate.

The first point to elucidate is, what proportion of this energy is emitted as radiation? Burton (1960) considers that this proportion is very large, and therefore luminescence quenching by oxygen has an important effect on the degree of radiation poisoning. Kohnlein and Muller (1962) noted that the aromatic amino acids require about 10–100 times more energy to form one free radical under the action of γ irradiation than do the aliphatic amino acids; this fact may be connected with large losses due to emission from the aromatic amino acids. Similar results have been obtained by Mochalkin (1965).

The radioluminescence yield of dry, deep-cooled trypsin determined by Augenstein *et al.* (1964*b*) is 0.097%; and according to our estimates and experimental results obtained for tryptophan solution but applied to proteins, this value does not exceed 0.26%. Because its yields are so low, radioluminescence does not appreciably effect the overall balance of absorbed energy or energy consumed on excitation. On the basis of the determined radioluminescence energy yields it may be said that in biopolymer solutions emission does not play an important part among the primary processes of dissipation of the absorbed radiation energy when the latter is sufficiently hard (above 1 MeV). Therefore most of the excitation energy will be used up in radiationless processes, among which attention has been focused [see the review by Barenboim *et al.* (1963*a*)] on the possibility of a singlet–triplet transition stimulated by oxygen. It is known that the effectiveness of the action of ionizing radiation generally increases in the presence of oxygen, which acts as a natural radiosensitizer; an analogous action is exerted by nitric oxide [see e.g., the review by Shchepot'eva *et al.* (1959)]. The sensitizing action has been related with the paramagnetism of O_2 and NO [Howard-Flanders (1957); Kihlman (1958); Howard-Flanders and Jockey (1960)]. Reid (1957) and later Mason (1959) connected hypothetically the sensitizing effect of oxygen with the fact that its paramagnetic molecule increases the probability of an intersystem crossing of the excited molecule to the triplet level. Analogous views about the role of excited states in the oxygen effect were voiced by Powers *et al.* (1960), Shekhtman (1960), and Zimmer (1961). In Mason's opinion the formation of a long-lived biradical in the triplet state (or the chromophore of a molecule in the triplet state) initiates denaturation of the protein; and according to Kasha (1960) it leads to the formation of peroxides. In photochemical reactions the formation of peroxides due to the effect of oxygen on the singlet–triplet transition has been observed experimentally in many cases [see the review by Karyakin (1961)].

Since the increased probability of a singlet–triplet transition in the presence of oxygen should lead to fluorescence quenching [Terenin (1947); Karyakin (1961)], the oxygen quenching of fluorescence induced by X or γ

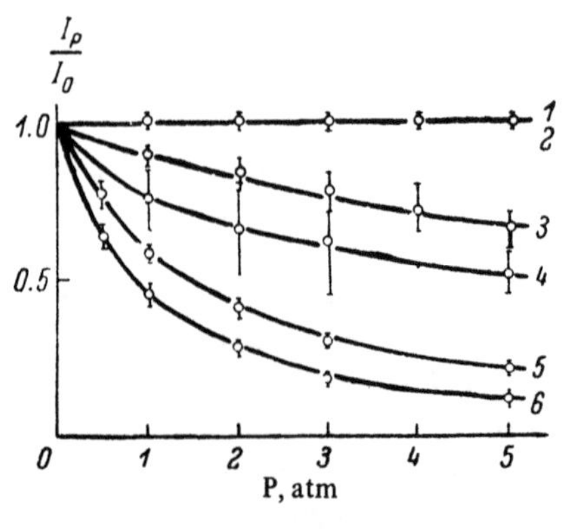

Fig. 3.2. Effect of various gases on the photofluorescence (PF) and radiofluorescence (RF) of tryptophan solutions. [0.01 *M* solution of tryptophan in 0.75 *N* alcoholic alkali. I_0 represents the fluorescence intensities in the presence of nitrogen (for PF) and helium (for RF) at 1 atm and I_p the intensities in he presence of nitrogen (helium) at higher pressures or in the presence of oxygen (nitric oxide) at any pressure]. (1,2) PF and RF in the presence of nitrogen and helium; (3) RF, nitric oxide quenching; (4) PF, nitric oxide quenching; (5) RF, oxygen quenching; and (6) PF, oxygen quenching.

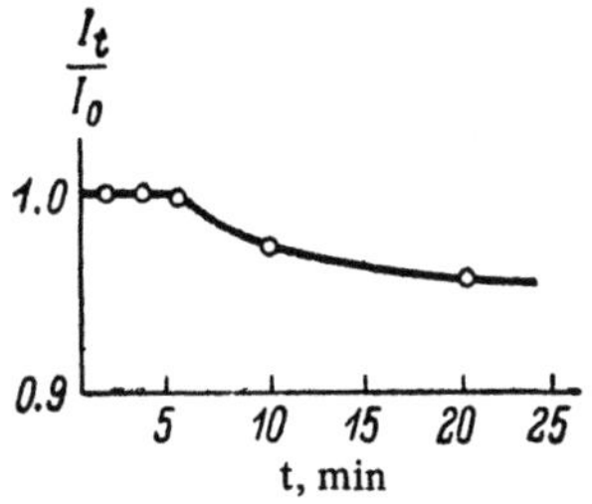

Fig. 3.3. Change in the radiofluorescence intensity during continuous X radiation at a constant pressure of oxygen (1 atm). I_t/I_0 is the ratio of the intensities at any time t to the initial intensity.

irradiation may serve as a method of confirming the hypothesis about the participation of excited states in the oxygen effect. Oxygen quenching of the radioluminescence of organic scintillators was measured by several authors [Kilin and Rozman (1963); Schram (1963); Birks (1964), Kilin *et al.* (1964)]. Oxygen quenching measurements for aromatic amino acids responsible for the emission from proteins were made by Barenboim (1963*a*, 1963*b*) and by Barenboim and Domanskii (1963). Figure 3.2 shows the effect of various concentrations of oxygen and other gases on the intensity of fluorescence excited by X rays. The observed quenching of the radiofluorescence is not related to oxidation which might give rise to nonfluorescent radiolysis products. One proof of this statement is the fact that the intensity of the emission excited by X rays remains constant throughout a time quite sufficient for the measurement (Fig. 3.3).

The quenching may be characterized by the dependence of the quenching coefficient K on the concentration C of the "quencher." In turn, $K = I_0/I_p$ where I_0 and I_p are the fluorescence intensities in an oxygen-free medium and in the presence of a certain concentration of oxygen (Fig. 3.4). A linear dependence of K on the oxygen pressure P allows us to introduce the concept of a rate

Table 3.2. Oxygen and Nitric Oxide Quenching of the Fluorescence of Aromatic Amino Acids and Casein

Fluorescent substance	Solvent	pH	Quenching gas	Type of excitation	Quenching rate K_T
Tryptophan solution	EtOH	9.5	NO	X rays	0.23
	EtOH	9.5	NO	UV	0.50
	EtOH	9.5	O_2	X rays	0.17
	EtOH	9.5	O_2	UV	0.28
	water	10.5	O_2	X rays	0.06
Tryptophan powder	–	–	O_2	X rays	0.02
Tyrosine solution	EtOH	9.5	O_2	X rays	0.06
Phenylalanine powder	–	–	O_2	X rays	0.03
Casein solution	water	10.5	O_2	X rays	0.04
Casein powder	–	–	O_2	X rays	0.03

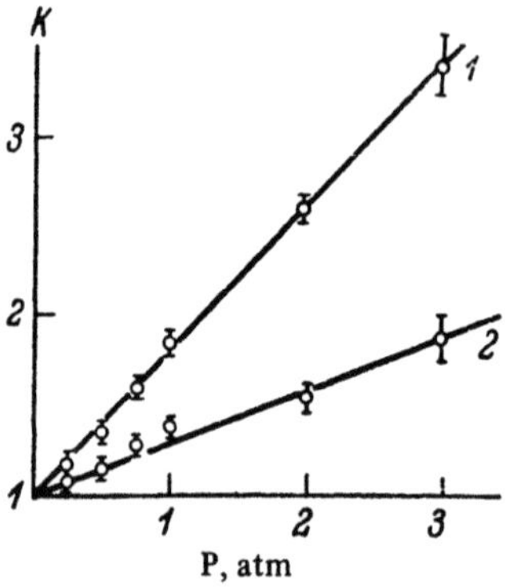

Fig. 3.4. Oxygen quenching of the X-ray fluorescence of tryptophan and tyrosine solutions in 0.075 *N* alcoholic alkali. (1) tryptophan, $5 \times 10^{-3} M$; (2) tyrosine, $5 \times 10^{-2} M$. The plots for solutions of tryptophan in the concentration range of $10^{-3}-5 \times 10^{-3} M$ lie within the reliability limits of line 1.

of quenching K_T, characterizing the change in K when the pressure of the quenching substance is changed by +1 atm:*

$$K = K_T P + 1$$

The values of K_T for the aromatic amino acids of certain proteins, nucleic acids, and guanine may be found in several publications [Barenboim (1963*a*, 1963*b*); Barenboim and Domanskii (1963)]; see also Table 3.2.

Table 3.2 and Figs. 3.2 and 3.4 show that at P corresponding to the atmospheric partial pressure of oxygen (154 mm) oxygen has practically no effect on fluorescence, although at higher pressures quenching is effective and is well described by a theory developed by Terenin (1947). Interpolation of the data to normal oxygen pressures in air and in the cell, and calculations based on experimental K_T values [see, for example, Barenboim and Domanskii (1963)], show that direct interaction of oxygen with molecules in the excited singlet level does not contribute seriously to radiation poisoning.

Thus the above-described experiments do not confirm Mason's hypothesis. Does this mean that excitation does not play any part in radiation effects? To answer this question a comparison is often drawn between the action of high-energy and UV radiation. For example, Lea (1959) compared qualitatively the action of these two types of radiation on the inactivation of viruses and bacteria and came to the conclusion that excitation has little effect in comparison with ionization. This view is shared by Gray (1955) and Bacq and Alexander (1961). However, the character of the excitation is not taken into account here: the significant formation (by "excitation") of triplet states due to the action of electrons, demonstrated experimentally, results in much greater chemical after-effects due to of radiation excitation (see above).

*It has been shown [Barenboim (1963*a*)] that under the conditions of the described experiments K_T depends only on the rate constants of fluorescence f, transition of the molecule into a triplet state under the action of oxygen t and interaction with external quenching substance k: $K_T = t/(f + K)$.

Another often-used proof for the lack of involvement of excitation in the radiation effect is that the threshold for inactivation of enzymes (trypsin, chymotrypsin) corresponds to particle (or photon) energies equal approximately to 10 eV and is close to the ionization potential of atoms in the protein molecule [Pollard *et al.* (1955); Setlow *et al.* (1959); Setlow (1958)]. However, such electrons may excite intensively the peptide linkages (see Section 3.1), and electron excitation of the singlet levels of peptide linkages may account for the effects of electrons with energies close to 10 eV. This is only a hypothesis, but one based on real physical laws of radioluminescence. Thus if the 10 eV inactivation threshold is to serve as a criterion of a predominant role of ionization we must first exclude this hypothesis by theory or experiment.

In recent years the relative roles of ionization and excitation were examined by Augenstein (1963) and by Augenstein *et al.* (1964*a*). The results obtained in a study of the influence of temperature at which the enzymes are irradiated on their yield of inactivation [Augenstein (1962)] cannot be interpreted correctly if only ionization is taken into account, but are well explained by effects associated with excitation [Augenstein (1963); Augenstein *et al.* (1964)]. The possible fate of the energy of this excitation is analyzed in detail in physical and chemical terms in the latter paper.

Thus the problem of the part played by excited states in the action of radiation on biological objects is still open.

Chapter 4

CHEMILUMINESCENCE OF CELLS AND ORGANISMS

Many biochemical reactions pass through exothermic stages which liberate energy. Such reactions as a rule involve molecules carrying excess energy, i.e., molecules in nonequilibrium excited states. The excitation energy is composed of the energy of the thermal effect and activation energy (Fig. 4.1). Deactivation of excited molecules occurs by various physical routes, and one of these is emission of the energy as light, viz. chemiluminescence.

The whole process of chemiluminescence can be considered [Vasil'ev (1965*a*)] as the formation of an excited product P* from the starting reactants K and M:

$$K + M \longrightarrow P^*,$$

followed by transition of the excited molecule P* to the ground state P with the emission of a photon

$$P^* \longrightarrow P + h\nu_p.$$

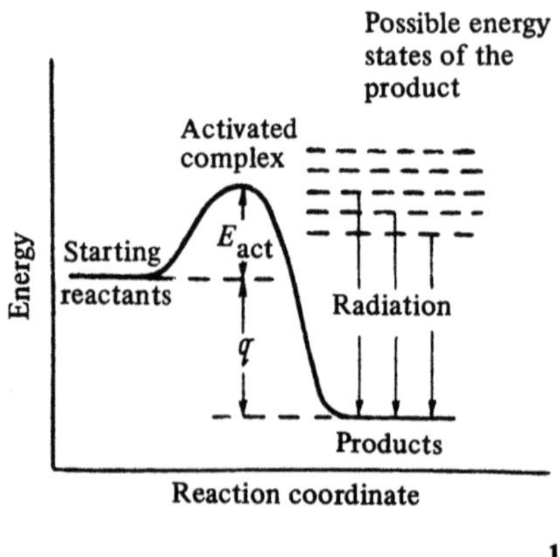

Fig. 4.1. Schematic illustration of the appearance of luminescence in an exothermic reaction [Vasil'ev (1964*b*)]. E_{act} is the activation energy and q is the heat of reaction.

The intensity of chemiluminescence can be expressed through the reaction rate if we can find the chemiluminescence quantum yield η_{cl} for every stage of the process [Vasil'ev (1965*a*)] :

$$\eta_{cl} = \eta_p \eta_p^*,$$

where η_p^* is the excitation yield, i.e., the ratio of excited molecules to the total number of molecules formed as a result of reaction, and η_p is the quantum yield of emission, i.e., the ratio of emitting molecules to the total number of excited molecules P*.

Then

$$I = \eta_{cl} w,$$

where I is the chemiluminescence intensity (the number of photons emitted per unit volume of the reacting system in unit time) and w is the reaction rate (the number of product molecules formed in unit time per unit volume).

The measurement of chemiluminescence intensity depends on the rate of the chemical process, on the probabilities of excitation and emission, and on the sensitivity of the light-receiving instrument.

Biochemical processes are known to occur in the liquid phase. The general physicochemical laws governing the chemiluminescence in liquid-phase reactions have been investigated in detail by Vasil'ev (1964*a*, 1964*b*; 1965*a*–*c*) and by Allabutaev *et al.* (1965). We shall now apply these laws in general form to the processes of light emission occurring in the course of biochemical reactions. These can be divided arbitrarily into three main types: (1) dark chemiluminescence, appearing as a side effect in metabolic reactions, in reactions intimately related with metabolism, and in destructive processes†; (2) functional chemiluminescence, associated with the presence of special systems whose function is production of light; and (3) induced chemiluminescence, appearing as a result of irradiation of the system.

Although all three types are in fact biological luminescence, the term bioluminescence is historically applied only the second type.

4.1. DARK CHEMILUMINESCENCE

Many investigations have shown that a great variety of living objects exhibit weak luminescence in the visible or ultraviolet part of the spectrum without actually containing any special light-producing system [see the reviews by Vladimirov and Litvin (1965); Zhuravlev *et al.* (1965); and Barenboim (1966)].

---

†The term "dark" is very arbitrary. To some extent it is opposed to the term "induced" chemiluminescence. However, the term "very weak" chemiluminescence is no better. Moreover, we can hardly call this emission "spontaneous," since it depends strongly on the external conditions (see Section 4.1).

Table 4.1. Dark Luminescence Intensity of Various Objects

Object	Background (pulses/min)	Emission intensity (pulses/min)	Temp. (°C)	Ref.
Mouse liver (opened abdominal cavity)	20–25	90–110	38	Tarusov *et al.* (1961*a*, 1961*b*); Tarusov *et al.* (1962)
Mouse liver homogenate	15–20	55–65	50	
Alcoholic extract of mouse liver lipids	50–60	100	60	
Ethereal extract of mouse liver lipids	50–60	280–340	60	
	50–60	69–72	30	
Mixture of alcoholic and ethereal extracts of mouse liver lipids with oxidizing fats	50–60	1500–1600	60	
Oleic acid	60–80	860–1200	60	
	22–27	39–52	30	
Human lipids:				
from the cerebral cortex	16–29	690–760*	60	Zhuravlev *et al.* (1965*a*)
from the heart	16–29	1320–1480*	60	
from the liver	16–29	180–230*	60	
from the kidneys	16–29	850–1015*	60	
Wheat stalk	1500–1800	5000*	43	Agaverdiev and Tarusov (1965)
Rat liver homogenate:				
in Ringer's solution	30	500	–	Vladimirov *et al.* (1962)
in sucrose	10	50	–	
Rat liver mitochondria	20	200	–	Vladimirov and L'vova (1964);
Liquid culture of malt yeast	10–12	46†	–	Konev *et al.* (1961*b*);
Synchronized culture of *Torula utilus*	–	200†	–	Konev *et al.* (1965*d*)
Frog skeletal muscle:				
resting	9–8	0	–	Troitskii *et al.* (1961)
fragmented	9–8	20†	–	
Contracting frog heart	9–8	35†	–	
Oxidative deamination of glycine under the action of H_2O_2	–	480,000‡	–	A.A. Gurvich *et al.* (1965)

When the temperature is not shown it was not reported in the publication in question. The emission intensities marked with an asterisk (*) have been recalculated to pulses/min·cm^2; those marked with a dagger (†) are expressed as percentage excess above the background; that with a double dagger (‡) is expressed as photons/min·cm^2.

Some idea of the dark chemiluminescence of various materials may be gathered from Table 4.1.

The intensity of the chemiluminescence is expressed in Table 4.1 as the number of pulses appearing per minute at the output of a photomultiplier recording the chemiluminescence. The resulting value may be 10–100 times smaller than the true intensity (photons/min·cm^2) because in this we are not considering the solid angle of light collection or the counting efficiency. Measurement of the chemiluminescence intensity in pulses/min·cm^2 without any corrections for the technical conditions of the measurement makes it impossible to compare the results reported by different authors. Evidently it is necessary to introduce corrections, e.g., by "tying" the measurements done on various objects to a standard quantity, say to the chemiluminescence intensity of oleic acid under standard conditions, or to a certain magnitude of the light flux of the scintillator with a radioisotope added.

By a historical coincidence, the studies of the dark chemiluminescence in the visible and UV regions of the spectrum were conducted practically independently of one another, in most cases on different objects and often even by different methods. Only recently have attempts been made to try and uncover common features in the emission mechanisms in these two regions of the spectrum.

Chemiluminescence in the Visible Region

This was first discovered in the higher plants by Colli and Facchini (1954), Colli *et al.* (1954), and Colli *et al.* (1955) and in mammals by Tarusov *et al.* (1961*a*). Further work showed that the main part of the overall emission in the visible region is due to reactions associated with nonenzymatic oxidation of lipids by oxygen [Zhuravlev *et al.* (1964, 1965*a–c*); Tarusov and Zhuravlev (1965)].*

Lipids are important biochemical cell components, comprising up to 40% by weight of the organic matter in the cell. The fatty acids forming from fats with the aid of lipases are second in importance to carbohydrates as energy sources in the organism.

The data on the chemiluminescence of some materials whose emission is ascribed to the oxidation of lipids are shown in Table 4.1. The emission from lipids stretches from the blue part of the spectrum to the infrared. In a particular case it was found that the chemiluminescence maxima of oleic acid and of liver homogenate in the region of 400 and 460 mμ coincide [Ivanov and Petrusevich (1965)]. The emission depends almost invariably on the temperature, pH of the

*A number of authors [Kaznacheev *et al.* (1964); Kaznacheev (1965); Kaznacheev *et al.* (1965*a–d*)] also discovered visible emission in blood cells and some other objects, using the photoelectric method. However, the molecular mechanisms of this emission were not analyzed.

medium, and concentration of oxygen [Veselovskii *et al.* (1963); Gasanov and Mamedov (1963)].

The Mechanism of Emission. It is known that free lipids can be intensely oxidized by oxygen. The fact that the oxidation of lipids in the cell is not direct but mainly enzymatic is due to the presence of antioxidants (oxidation inhibitors), which retard the chain oxidations of lipids necessary for the spontaneous oxidation reactions [Zhuravlev *et al.* (1961)]. The cell antioxidants are lipidic in nature [Zhuravlev (1959); Witting (1965)] and include some phospholipids such as tocopherols, lecithins, etc.

The reactions between cell antioxidants and the free peroxide radicals due to the nonenzymatic oxidation of fats were first regarded as one of the main sources of emission [Zhuravlev *et al.* (1961); Zhuravlev and Veselovskii (1963); Popov and Tarusov (1963, 1964)]. It was assumed that the photon is emitted from an excited inhibitor molecule InH, which as a result transforms into a radical incapable of initiating a chain reaction.

$$R^{\cdot} + InH \longrightarrow RH + In^{*} \longrightarrow RH + In^{\cdot} + h\nu,$$
$$RO_2^{\cdot} + InH \longrightarrow ROOH + In^{*} \longrightarrow ROOH + In^{\cdot} + h\nu.$$

Later, under the influence of ideas developed by Vasil'ev (1964*a*; 1965*a*), the same authors [Zhuravlev *et al.* (1964, 1965*a–c*); Tarusov and Zhuravlev (1965)] proposed for the chemiluminescence of lipids in biological systems, a mechanism that took into account the possibility of the transfer of the excitation energy to an activator substance not involved in the reaction but capable of emitting the photons. Such a substance is, for example, ergosterol: when this compound is added to oleic acid, it does not alter the latter's rate of autoxidation but enhances the luminescence [Zhuravlev *et al.* (1965*b*)]. The part of activators may also be played by certain oxidation inhibitors [Zhuravlev *et al.* 1965*c*)]. The general scheme of the reactions examined in the above-cited studies is as follows:

1. Chain oxidation of fats

$$R^{\cdot} + O_2 \longrightarrow RO_2^{\cdot},$$
$$RO_2^{\cdot} + RH \longrightarrow ROOH + R^{\cdot}.$$

2. Recombination, with the formation of excited states

$$RO_2^{\cdot} + RO_2^{\cdot} \begin{cases} \nearrow [P]^{*} + O_2 \\ \searrow R_2 + [O_2]^{*}. \end{cases}$$

3. Luminescence in the absence of activator

$$[P]^{*} \longrightarrow P + h\nu.$$

4. Luminescence with an activator (A)

$$[P]^{*} + A \longrightarrow P + [A]^{*} \longrightarrow P + A + h\nu.$$

5. Luminescence during reaction with InH:

$$\left.\begin{matrix} R^{\cdot} \\ RO_2^{\cdot} \end{matrix}\right\} + InH \longrightarrow \left.\begin{matrix} RH \\ ROOH \end{matrix}\right\} + In^{\cdot},$$

$$In^{\cdot} + O_2 \longrightarrow InO_2^{\cdot},$$

$$In^{\cdot} + RO_2^{\cdot} \longrightarrow [ROOIn]^* \longrightarrow h\nu,$$

$$InO_2^{\cdot} + RO_2^{\cdot} \longrightarrow [P]^* \longrightarrow h\nu.$$

According to these authors' hypothesis, the antioxidants not only interrupt the chain reaction in lipids but also protect the activator from peroxide radicals. However, it seems to us that despite its apparent clarity this scheme does not solve the problem of the mechanism of chemiluminescence in the visible region of the spectrum, but only indicates some possible reaction paths.

It may be noted that cell destruction during electrolysis also leads to chemiluminescence which is associated with a reaction between the oxidation inhibitors and radicals formed in the electrolysis [Pyatenko and Tarusov, (1964)]. However, the mechanism of the emission in this case is evidently not necessarily connected with inhibitors: luminescence during electrolysis appears in a great variety of organic acids, salts, etc. [see the review by Vasil'ev (1965*b*)], and in various aromatic compounds [Hercules (1964)]. In particular, intense emission appears in the electrolysis of tyrosine, phenylalanine, histidine, and tryptophan solutions [Petrusevich and Ivanov (1965)].

Vladimirov *et al.* (1962) proposed that the basic feature underlying the mechanism of the dark luminescence may be other biochemical reactions, e.g., those connected with enzymatic processes. Thus, in their opinion, the enzymatic decomposition of peroxides formed in the course of cell metabolism, and especially the decomposition of H_2O_2 by catalase and peroxidase, is accompanied by luminescence. On the other hand, according to Likhtenshtein *et al.* (1965), the emission in this reaction is due not to the enzymatic activity of catalase but to oxidative interactions between H_2O_2 and certain groups of the apoenzyme. This, too, may be the reason for the emission in the interaction of peroxidase with H_2O_2 described by Cormier and Dure [Cormier and Dure (1963); Dure and Cormier (1963*a*, 1963*b*, 1964)], which in Vladimirov's [Vladimirov *et al.* (1962)] opinion arises as a result of an enzymatic process.

It may be proposed that in the decompositions of peroxides the rate is determined by catalase and peroxidase, and the decomposition products in turn enter into reactions with the protein part of the enzymes. The action of catalase and peroxidase is therefore associated with luminescence. Analogous emission during the decomposition of H_2O_2 in various enzyme systems has been observed by Konev *et al.* (1961*b*) and by Sapezhinskii *et al.* (1965*a*), and in rat liver homogenates by Novikov (unpublished work; see Fig. 4.2).

Gusar *et al.* (1965) observed emission as a result of pure enzyme–substrate interactions in sucrose–invertase, β fructofuranosidase, and casein–trypsin

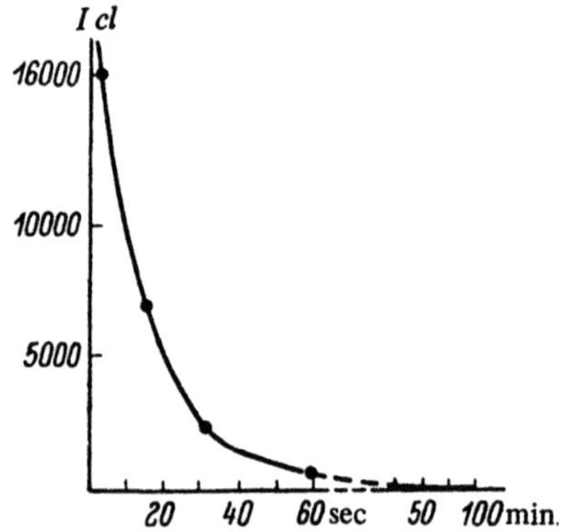

Fig. 4.2. Chemiluminescence from rat liver homogenate after addition of H_2O_2. I_{cl} is the intensity in pulses/sec. The abscissa represents the time from the addition of peroxide.

systems. It is still difficult to formulate a definite mechanism for such luminescence. Trypsin is known to cleave peptide bonds near the basic amino acids arginine and lysine [Bresler (1963)]:

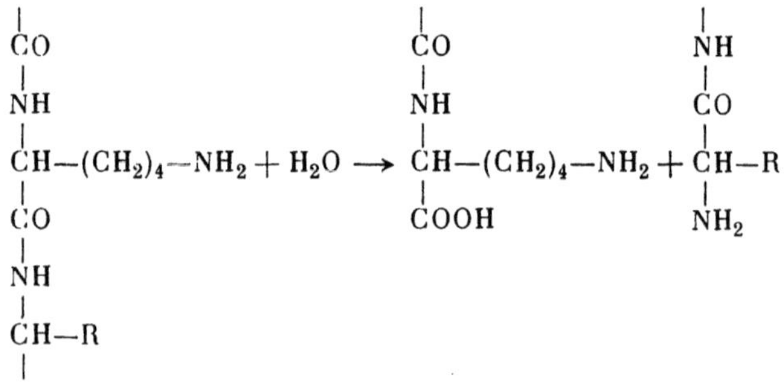

The peptide link would release about 69.7 kcal/mole [Pullman and Pullman (1963)], formation of a carboxyl group takes about 90 kcal/mole [Vedeneev *et al.* (1962)], and the activation energy for trypsin hydrolysis of casein is about 12 kcal/mole [Bull (1947)]. In contrast, the energy of the luminescence photons is much higher than the arithmetical sum of these – it is about 60 kcal/mole.

The details of Gusar's method are unknown, but if the reaction occurred in the presence of atmospheric oxygen then it is possible that the emission was due to interaction of oxygen with the hydrolysis products. According to A. A. Gurvich (1965), UV emission of several enzyme systems, which will be discussed later, appeared only in the presence of oxygen.

We are still unable to cite reliable examples of luminescence of enzyme systems in the visible region appearing solely as a result of an enzyme–substrate reaction. This may be because enzymes reduce sharply the activation energy, E of the reaction. Thus, in the nonenzymic decomposition of H_2O_2, E = 18 kcal/mole, while in the presence of liver catalase E = 5.5 kcal/mole [Bull (1947)].

In turn, the magnitude of E influences the magnitude of the excitation energy (Fig. 4.1) so that in many cases the action of enzymes may destroy the chemiluminescence visible in the same reactions occurring nonenzymatically.

Thus the main part of the dark chemiluminescence in the visible region – up to 75% of the total emission from native tissue homogenates [Tarusov and Zhuravlev (1965)] – is due to oxidation of lipids, though in principle the oxidation of other substances may also make a contribution: possible examples are oxidation of polyphenols, which chemiluminesced in a model reaction [Slawinska and Slawinski (1964)].

Localization of the Emission. Apart from serving as one of the energy sources in the cell, lipids participate in the formation of a number of structures: the cell wall, mitochondrial membranes, the membrane of the endoplasmic reticulum, etc. Is the luminescence associated with all lipids of the cell, or is it due predominantly to lipids localized in specific organelles, just as in the cell the UV fluorescence is due predominantly to mitochondrial proteins (Section 1.6)? At the moment we simply do not know. Petrusevich and Konoplyannikov (1965) proposed on indirect evidence that the lipids responsible for luminescence are those in the cell wall. Similar results were obtained earlier by Stauff and Reske (1964), though these authors did not discuss the localization of the emission in the cell. Vladimirov and L'vova (1964) observed emission from intact mitochondria of rat liver, which they too associated with oxidation of lipids [Vladimirov and L'vova (1965*a*)]. In principle, it may be supposed that the mechanism of the emission from mitochondrial lipids is similar to that examined above, and that the luminescence is due both to the structural lipids and those contained in the matrix. In both cases we can envisage nonenzymatic interaction between lipids and oxygen, formation of peroxides, and participation of oxidation inhibitors–phospholipids, especially lecithin and α-tocopherol, which are present in mitochondria [Lehninger (1964)]. According to Vladimirov and L'vova (1964) the chemiluminescence intensity of mitochondria is associated with oxidative phosphorylation. Exclusion from the medium in which the mitochondria were incubated of any component of oxidative phosphorylation (ATP, oxidation substrates, oxygen, etc.) led to reduction or disappearance of the emission. The authors propose that oxidative phosphorylation provides energy for the oxidation of lipids (in particular for the oxidation of fatty acids).

It may be postulated that the mechanical and chemical changes of mitochondria associated with oxidative phosphorylation influence the chemiluminescence of lipids via changes in the conformation of lipoprotein complexes [e.g., Lehninger (1964); see also Section 1.6].

Mitogenetic Radiation (Dark Chemiluminescence in the UV)

Toward the end of the 1920's A. G. Gurvich discovered that dividing cells of onion root, regenerating cornea of frog's eye, dividing cells of malt yeast, blood cells, and certain other objects luminesce weakly in the region of

190–325 mμ [A. G. Gurvich and L. D. Gurvich (1934, 1947, 1948)]. This emission was perceived by biological detectors: onion root cells or yeast cells. Stimulation of mitosis constituted a measure of the emission, and hence the name "mitogenetic".*

Many authors [see the reviews by Moiseeva (1960*b*)] consider that the method of biological detectors is too underdeveloped and too susceptible to the influence of extraneous factors, so that the very existence of the phenomenon is uncertain. However, UV emission of cells was detected with gas-discharge counters as early as the mid-thirties [Rodionov and Frank (1934); Barth (1936); Siebert and Seffert (1936*a*, 1936*b*); Audubert (1938, 1939)]. In recent years mitogenetic radiation has been studied by many authors using counters or photon-counter photomultipliers [Becher (1957); Vladimirov and Litvin (1959*b*, 1965); Konev *et al.* (1961*b*); Troitskii *et al.* (1961); Vladimirov *et al.* (1962); A. A. Gurvich *et al.* (1965); Konev *et al.* (1965*a*)] and has been recorded photographically [Sitko (1960)].

Mechanism of the Emission. The sources of the UV emission from cells have not as yet been precisely defined. Very general ideas were put forth by Reinstoff (1960), who proposed that the main sources were oxidation, proteolysis, and glycolysis. According to A. G. Gurvich and his associates [see, e.g., A. G. Gurvich (1959); A. A. Gurvich (1965); A. A. Gurvich and Eremeev (1965)] far-reaching enzymatic cleavage of the substrate is statistically rare, i.e., small free radicals are formed: hydroxyls, carbonyl groups, amino or imino groups, atoms of oxygen, hydrogen, and so on. Recombination of these radicals supplies energy which is absorbed and then reemitted by the surrounding molecules. Several experimental works and calculations of the energy balance support these ideas [A. A. Gurvich and Eremeev (1965)]. Since UV emission was observed mainly in the presence of oxygen [A. A. Gurvich (1965)] it may be due to reactions between oxygen and radicals formed as a result of enzymatic cleavage processes.

In a number of cases it may be assumed that the same reactions are responsible for luminescence in the visible region of the spectrum and in the ultraviolet. A maximum in the blue end of the continuous spectrum may indicate that part of the energy is emitted in the UV. In fact, studying the luminescence of the root system of barley Konev and Katibnikov (1965) found a strong maximum at about 480 mμ and weak UV emission. Any process enhancing the overall intensity of the dark luminescence will also increase the UV component. This may occur, for example, during mitosis. E. M. Brumberg *et al.* (1961*c*) did indeed find in this case an increase in the intensity of cell photoluminescence. In turn, according to Vasil'ev's (1965*a*) data, a rise in the photoluminescence yield enhances chemiluminescence proportionately. This is

*At present this term is often used to denote the chemiluminescence emission of biological objects in the UV, independently of the detector used.

not to say that in other cases the visible and the UV luminescence may not be due to different reactions or to different mechanisms in reactions of the same type. For example, A. A. Gurvich *et al.* (1965) observed visible and UV emission in the oxidative deamination of glycine in the presence of H_2O_2 and proposed that different mechanisms are responsible for these two forms of emission.

In certain cases the sources of visible and UV chemiluminescence may be related reactions: if the oxidation of lipids leads to visible luminescence, the resulting aldehydes and ketones may then chemiluminesce in the near UV when they in turn become further oxidized [Russell (1957); Berezin *et al.* (1962)].

The Biological Role of Mitogenetic Radiation. A. G. Gurvich and L. D. Gurvich (1948) [see also Zalkind (1937*a*, 1937*b*, 1964)] described some observations demonstrating that mitogenetic emission is a necessary condition of mitosis, stimulates mitotic activity, induces reversible morphological changes in cells, modifies cell permeability, and so on.

In spite of the critics of biological detectors [Moiseeva (1960*a*, 1960*b*); Hollander (1960)], several workers [A. A. Gurvich (1960, 1961, 1965); A. A. Gurvich *et al.* (1965); Eremeev (1961)] again confirm the influence of the UV emission of inducers on the mitotic activity of the detectors (onion root cells, yeast cultures).

The mutual influence of growing cells divided by a quartz partition transparent to the ultraviolet, described already by A. G. Gurvich, found confirmation in the investigations of Kaznacheev *et al.* (1964) on fibroblast tissue cultures and in Kulin's work (1960, 1965). Using photomultipliers, Konev *et al.* (1965*a*) found that the maximum intensity of the emission of a synchronized culture of *Torula utilus* (λ_{max} = 330 mμ) corresponds to the premitotic period.

An indirect confirmation of the effect of mitogenetic radiation on living cells is also provided by model experiments with UV lamps. Under these conditions it was found that light similar in its characteristics to the "mitogenetic" radiation activates such nonphotobiological processes as respiration of yeasts [Galston (1950)] segmentation of sea urchin eggs [Kimball (1955)], emergence of bacteriophage from cells [Vladimirov *et al.* (1962)], photosynthesis of a green leaf previously kept in the dark, and mitosis of the epithelium of cooled cornea of frog's eye [A. A. Gurvich (1964)]. Using a gasometric method, Konev and Lyskova (1964) showed that a light flux of 10^8 photons/cm^2·sec at a wavelength of 254 mμ stimulates the rate of glycolysis in yeasts, and a flux of 10^5-10^6 photons/cm^2·sec at 280 mμ accelerates the occurrence of the first wave of budding in a synchronized yeast culture.

Thus in many cases weak UV irradiation may affect various aspects of the life activity of the cell. Does this mean that the UV produced by a living system plays in it some special part? We believe that any such conclusion would be premature. There is nothing surprising in the fact that the cell is affected by this highly photochemically active form of radiation. However, we do not know of any experimental proof that UV emission and absorption in the course of chemiluminescence is one of the controlling mechanisms in the cell rather than a statistical fluctuation, a kind of interference, exerting an infrequent, random influence on the system or the reaction in question. For example, the simple observation that the emission from some cells affects other cells does not necessarily mean that such an effect has any biological significance. In exactly the same way, the theoretical possibility of the transmission of a large flow of information on light frequencies obviously does not prove that in living systems the information necessary for the initiation and control of various cytochemical processes is obtained in just this way, as believed by Kaznacheev and his associates [Kaznacheev *et al.* (1964); Kaznacheev (1965)]. There are still no experimental grounds for any connection between mitogenetic radiation and hypothetical nonequilibrium excited molecular constellations [A. A. Gurvich *et al.* (1965)]. There is still no proof that the UV emission appearing in chemiluminescence contributes to the overall energetics of the cell. Very frequently the results of measurements are very little above the background of the light receivers [Vladimirov and Litvin (1959*b*); Troitskii *et al.* (1961)]. In other words, the quantitative material on the interaction of cell radiation is limited, and as a rule there is no statistical treatment of objectively selected characteristics [Kaznacheev *et al.* (1964); Kulin (1965)].

In our opinion the solution of the whole problem of mitogenetic radiation is in its initial phase. The recent work has demonstrated the existence of cell UV emission but has contributed little to an understanding of its significance. It may therefore be postulated that this emission is often a special case of dark chemiluminescence, i.e., a "pointless" loss of energy, especially as the emission mechanisms are supposed to be the same in a number of cases. In other cases it may be expected that the UV radiation exerts a random and nonspecific action on the living system. Are there any other possibilities? Improvement of the techniques of recording very weak UV emission, successful development of the theoretical side of chemiluminescence, the recent successes of molecular biology, and the renewed interest of investigators in mitogenetic radiation allows us to hope that the problem will eventually be solved.

Utilization of Dark Chemiluminescence for the Study of Life Processes

Study of chemiluminescence makes it possible to obtain various kinds of information about the mechanisms of chemical reactions. One can thus determine the number of stages in a complex reaction, the rate constants and activation energies, rates of disappearance of free radicals, the part played in the reaction by excited molecules, and so on [Vasil'ev (1965*a*, 1965*b*)]. The chemiluminescence parameters are connected with certain characteristics of the light-emitting molecules – their structure, the probabilities and energies of electronic transitions, and donor–acceptor properties. However, before passing to such molecular and quantum-mechanical characteristics in living systems it must be established which molecules are in fact responsible for the emission of the photon. The reaction mechanisms leading to the formation of excited states and the light-emitting substances in the living cell are still unknown. Their identification is the main field of study in dark chemiluminescence, since this information is necessary for a full interpretation of the observed phenomena. Nevertheless, dark chemiluminescence has proved useful in cytological and physiological work. It allows some conclusions to be drawn about the general level of the physiological activity of the cell [Vladimirov and L'vova, (1964, 1965*a*); Mamedov (1965)] changes in the state of some functional systems – the nervous system, skeletal and cardiac muscles – [A.A. Gurvich (1964)], the life activity of certain tissues [Popov and Tarusov (1963, 1964)], mitotic activity of the cell [Gawronski (1961)], heat and cold stability of various plant cultures [Gasanov *et al.* (1963*a*, 1963*b*); Agaverdiev and Tarusov (1965*a*, 1965*b*)], antitumoral and radioprotective properties of some substances [Ismailova *et al.* (1965); Mkrtchyan *et al.* (1965); Petrusevich *et al.* (1965); Tarusov *et al.* (1965); Tkhor and Kozlov (1965), and others].

By measuring the chemiluminescence intensity in kinetic experiments *in vitro* it was possible in certain cases to determine the number of stages in a complex reaction, the rate constants and activation energies of these stages [Popov and Tarusov (1963, 1964)], the dependence of the rate of oxidation on oxygen concentration [Popov and Tarusov (1963)], and the activity of peroxidase [Nilsson (1964)]. In the electrolysis of aromatic amino acids chemiluminescence offers a convenient way of investigating free-radical states [Klipson *et al.* (1965)].

4.2. FUNCTIONAL CHEMILUMINESCENCE (BIOLUMINESCENCE)

Chemiluminescence more intense than dark chemiluminescence is fairly often found in nature. This type of chemiluminescence is the main result of certain specialized reactions within the organism.

Numerous investigations have shown that the visible glow from living organisms – bioluminescence – is by no means rare. It is found among bacteria, fungi, radiolaria, sponges, crustaceans, insects, fishes, and many other organisms. Harvey (1940, 1952) considers that about 50% of all species have luminescent forms. Such forms are found most often among marine organisms. Bioluminescence has never been observed in the higher plants, birds, or mammals. In spite of this wide distribution of bioluminescence and the age-old interest in the subject, the physicochemical processes underlying the phenomenon are still incompletely understood.

According to Seliger and McElroy (1960), bioluminescence is based on enzyme-catalyzed oxidation in which the excess energy is emitted in the form of visible light. The oxidizing substance is usually luciferin, and the enzyme is

Table 4.2. Types of Bioluminescent Systems

Organisms	Reaction system	Reactions causing luminescence	Max. of the emission (mμ)
Bacteria (*Photobacterium fischeri* and others)	Pyridine nucleotide	1) $NAD\text{-}H_2 + FMN \underset{}{\overset{NAD\text{-}H_2-\text{oxidoreductase}}{\rightleftarrows}} FMN\text{-}H_2 + NAD$ 2) $FMN\text{-}H_2 + RCHO + O_2 \xrightarrow{E}$ light	465–495
Fungi (*Omphali flavida* and others)	Pyridine nucleotide	1) $NAD\text{-}H_2 + L \overset{NAD\text{-}H_2\text{-oxidoreductase}}{\rightleftarrows} LH_2 + NAD$ 2) $LH_2 + O_2 \xrightarrow{E}$ light	530
Insects (*Photinus pyralis* and others)	Adenine nucleotide	$LH_2 + ATP + O_2 \xrightarrow[Mg^{2+}]{E}$ light	562
Corals (*Renilla reniformis*)	Adenine nucleotide	$LH_2 + ADP + O_2 \xrightarrow{E}$ light	Blue region
Crustacea (*Cypridina hilgendorfii* and others)	Enzyme substrate	$LH_2 + O_2 \xrightarrow{E}$ light	460
Dinoflagellate (*Gonyaulax polyhedra*)	Enzyme substrate		470
Worms (*Odontosyllis enopla* and others)	Enzyme substrate		510
Mollusks (*Pholas dactylus*)	Enzyme substrate		480
Fishes (*Parapriacanthus beryciformis* and others)	Enzyme substrate		460
Enteropneusta (*Balanoglossus biminensis*)	Peroxidase	$LH_2 + H_2O_2 \xrightarrow{E}$ light	Blue region
Jellyfish (*Aequorea aequorea* and others)	—	Protein $\xrightarrow[Ca^{2+}]{}$ light	460 and 510

LH_2 stands for reduced luciferin, E for luciferase. The table is compiled mainly from McElroy and Seliger (1962*b*, 1962*d*) as well as from Cormier and Totter (1964) and Chase (1964).

luciferase. These terms are traditional and very arbitrary, and the compositions of the components are different for most organisms. The specific mechanisms of the reactions in question are correspondingly different.

In contrast to dark chemiluminescence, bioluminescence has long been studied, and much has been published on the subject, including several detailed reviews on all aspects of the phenomenon. Thus Harvey (1952) gives a detailed description of the ecology of luminescent organisms, comparative anatomy of light-producing structures, and physiology of the process. Some anatomical, physiological, and physicochemical aspects are examined by Johnson and others [Johnson (1955); Johnson *et al.* (1959); Johnson *et al.* (1961); Nicol (1960, 1962); and Chase (1964)]. McElroy (1960, 1961) and McElroy and Seliger (1961, 1963) consider in detail the biochemical aspect. In the extensive review by Cormier and Totter (1964) an attempt is made to classify luminescent organisms according to the type of biochemical reaction responsible for the emission. Soviet literature contains few such reviews [Grinberg (1920); Sveshnikov (1941); Zhuravlev *et al.* (1965*c*); Barenboim (1966)]. In the last two articles bioluminescence is considered together with other forms of the chemiluminescence of biological objects. The role of bioluminescence of marine organisms in the luminescence of sea water is discussed at length by Tarasov (1956). There are also a number of shorter, "popular" reviews [McElroy and Seliger (1962*a*); Nicol (1963); Gauroy and Grive (1964); Korten (1964); Shimomura (1964)].

Since the subject has thus been discussed from various aspects, we shall content ourselves in this book with a summarizing compilation (Table 4.2) and consider as examples the bioluminescence systems of the firefly, which has been most studied, of bacteria, whose luminescence is used widely as a test for various investigations, and of the jellyfish, whose bioluminescence depends on an original and little-studied mechanism.

Bioluminescence of the Firefly

The mechanism of the glow of insects has not yet been fully explained, in spite of a large number of studies [see, e.g., Rhodes and McElroy (1958*a*); McElroy (1960, 1962, 1963*a*, 1963*b*); McElroy and Seliger (1961, 1962*a*–*c*); De Luca and McElroy (1965)]. The reaction involves luciferin (LH_2), luciferase (E), ATP, magnesium ions, and oxygen. McElroy and Seliger (1963) suggested the following series of reactions for the luminescence of insects:

$$E + LH_2 + ATP \overset{Mg^{++}}{\rightleftharpoons} [E{-}LH_2{-}AMP] + PP,$$

$$[E{-}LH_2{-}AMP] + O_2 \longrightarrow [E{-}LO{-}AMP]^* + H_2O,$$

$$[E{-}LO{-}AMP]^* \longrightarrow [E{-}LO{-}AMP] + \text{Light}$$

where PP is pyrophosphate and LO is oxidized luciferin. Subsequently, the free pyrophosphate splits the inactive complex, and luciferase and ATP again enter the reaction:

$$[E{-}LO{-}AMP] + PP \xrightarrow{Mg^{++}} LO + E{-}ATP.$$

However, recently De Luca and McElroy (1965) found that luciferase itself can hydrolyze the inactive complex, splitting out AMP, appearing in the reaction as a hydrolase.

The luciferin of the firefly, obtained in crystalline form by Bitler and McElroy (1957), was subsequently synthesized by White *et al.* (1961) by condensation of 2-cyano-6-hydroxybenzthiazole and D-cysteine. Reduced luciferin has the following structure:

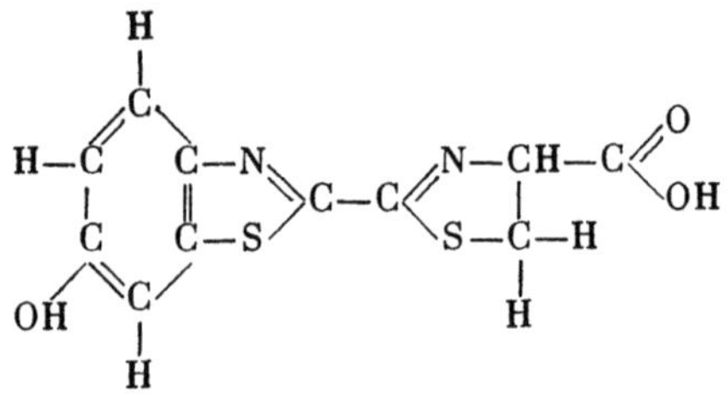

Various analogs of luciferin were synthesized in recent years [White *et al.* (1965)].

To demonstrate the difference between the composition of luciferins from various organisms, we give below the structural formula of a second luciferin (of the two elucidated up to now) obtained from the crustacean *Cypridina** by Shimomura *et al.* (1957) and by Hirata *et al.* (1959):

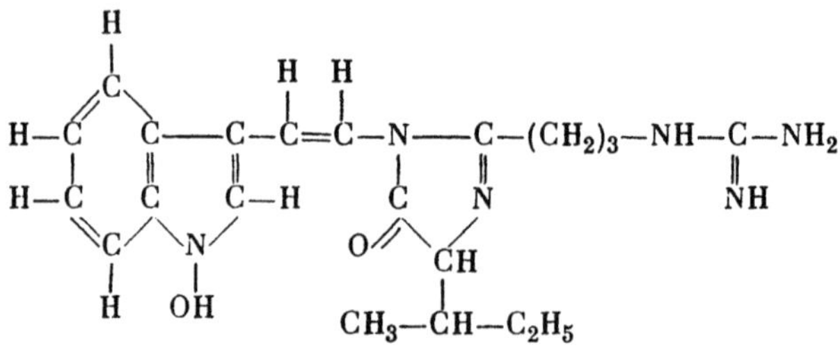

Firefly luciferin is capable of autoxidation (without chemiluminescence) and yellow-green fluorescence on absorption in the region of 395 mμ [McElroy and Seliger (1961)].

Firefly luciferase has been crystallized in pure form by Green and McElroy (1956). The method is described in detail by McElroy (1963*b*).

*The luciferin and luciferase of *Cypridina* are used as a convenient model for studying the action of irradiation and protective substances [Lohmann *et al.* (1963, 1964)]. Moreover, it has been shown that luciferin exhibits anticarcinogenic activity [Shinazo (1964)].

The exact structure of firefly luciferase is still unknown. It is only supposed that it is a protein belonging to the euglobulins, which, according to data obtained by electrophoresis, ultracentrifugation, and tests of enzyme activity, is homogenous after fivefold recrystallization [McElroy (1963*b*)]. Luciferase contains seven SH groups, only two of which affect the catalytic activity of the enzyme [De Luca *et al.* (1964)]. The isoelectric point of luciferase is around pH 6.2–6.3, and its sedimentation constant, S_{25}, in 0.15*M* NaCl is 5.6 S. The molecular weight, determined from the sedimentation constant, is around 10^5 [McElroy and Seliger (1963)].

The enzyme exhibits maximum activity at 25–27°C and pH 7.8 [McElroy (1963*b*)]. The activity is strongly inhibited by amines, copper, and *p*-chloromercuribenzoate, but does not change in the presence of arsenic, potassium cyanide, and azides [McElroy (1963*b*)].

Study of the bioluminescence of certain species of firefly revealed species-specificity, evidenced by the different production of emission spectra [Seliger and McElroy (1964)] and by differences in the latent period and duration of the flash in response to an electrical stimulus [J. B. Buck and Case (1961)]. The differences between the luminescence spectra of various species of the firefly were ascribed to species-specificity of luciferase, i.e., to different configurations of the enzyme molecules. It is known that the glow is due to the complex (E–LO–AMP): at the same time it has been found that the luciferins of these species give the same chromatograms, absorption spectra, and emission spectra, so that they evidently have the same structure.

The existence of species-specificity in the bioluminescence of insects may be related to a biological function, i.e., with recognition of the opposite sex by individuals belonging to the same species.

The firefly system is the most effective of all the known chemiluminescence systems. Seliger and McElroy (1960) showed that at neutral and alkaline pH the luminescence quantum yield of the firefly is close to unity.

The luminescence intensity depends linearly on the concentration of ATP, and this served as a basis for a chemiluminescence method for the determination of ATP in a wide range of systems. The method involves introduction of a firefly extract containing luciferin and luciferase into the sample. This procedure may allow a determination of as little as 10^{-11} g of ATP per ml of solution [McElroy and Strehler (1954); Cormier (1961); Tumerman (1962); Masterson *et al.* (1964); Prydz and Froeholm (1964); Tumerman and Fedorovich (1965)].

The Bioluminescence of Bacteria

Some bacterial species exhibit continuous emission with maxima between 465 and 495 mμ (Table 4.2). This phenomenon is associated with respiration [Harvey (1952); McElroy and Seliger (1962*a*), Chumakova (1963*a*, 1963*b*)]. McElroy and Seliger give a scheme (Fig. 4.3) which shows that the light-producing reaction branches off from the overall electron-transfer process of the

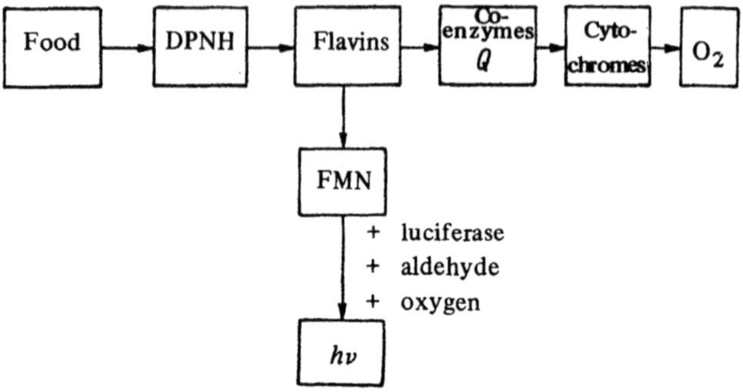

Fig. 4.3. Scheme showing the postulated point of branching of the light-producing reaction from the overall electron-transfer process in bacteria [McElroy and Seliger (1962*a*)].

respiratory chain; the branching takes place above the level of dehydrogenases and the Krebs cycle [Chumakova (1963*b*); Chumakova and Egorova (1964*a*, 1964*b*)]. Gibson and Hastings (1962), McElroy and Seliger (1963), and Chumakova (1963*b*) recently proposed that the luminescence of bacteria appears in the luciferase-catalyzed oxidation of a "reduced flavin mononucleotide–aldehyde complex," which is regarded as bacterial luciferin. It has been shown in extracts from bacteria [McElroy (1961)] that a whole range of the higher (C_6–C_{18}) aliphatic aldehydes can participate in the light-producing reaction.

Bacterial luciferase was obtained in crystalline form only recently [Kuwabara *et al.* (1964, 1965); Hastings *et al.* (1965)]. It is a protein with a sedimentation constant S_{20} = 5.15 S and a molecular weight of 7.6×10^4. The chemiluminescence quantum yield calculated from the molecular weight is about 0.28 per luciferase molecule.

In the present authors' laboratory S. N. Borkhsenius recently attempted to isolate luciferase from *Photobacterium* sp. obtained by R. I. Chumakova from the Pacific (36th voyage of the research vessel "Vityaz").

The bacterial mass was obtained in a special fermenter. The cells were stored in the frozen state and were opened up by grinding with alumina in a buffer containing β-mercaptoethanol. The cell extracts were freed from nucleic acids with streptomycin and were then fractionated with ammonium sulfate and by chromatography on DEAE cellulose, using mainly the procedure employed by Kuwabara *et al.* (1965) and Hastings *et al.* (1965) to isolate luciferase from *Photobacterium fischeri.* After chromatography on DEAE cellulose, one of the fractions exhibited luciferase and NAD–H_2–oxidoreductase activity, had an absorption spectrum with maxima at 280 and 407 mμ and was homogeneous when investigated in an analytical centrifuge, with a sedimentation coefficient S_{20} = 3.95 S. Crystallization of this material gave two types of crystals (Fig. 4.4): convex polyhedra with 12 faces and hexagonal prisms with 8 faces.

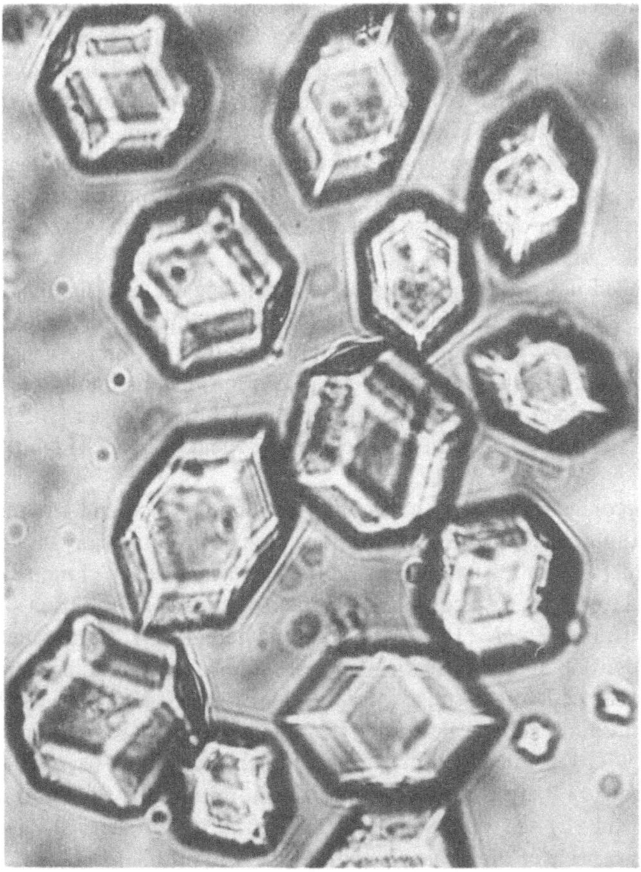

Fig. 4.4. Crystals of luciferase from *Photobacterium* species. Optical microscope (X 400). The mean crystal diameter is 10 mμ.

The dodecahedral crystals are of the same form as the luciferase crystals obtained by Kuwabara *et al.* (1965). It is worth noting that Hastings' crystalline luciferase preparations [Hastings *et al.* (1965)] from *Photobacterium fischeri* also exhibited NAD–H_2–oxidoreductase activity.

It is believed that one of the above crystal types was formed by luciferase and the other by NAD–H_2–oxidoreductase. Work on the separation of these fractions by chromatography on carboxymethylcellulose, hydroxyapatite, and other ion-exchangers is still in progress.

The formation of a single quantum of emitted light in bacteria involves 20 long-chain aldehyde molecules, about 2800 $NADH_2$ (or DPNH) molecules [Johnson *et al.* (1961)], and 150 molecules of FMN [Terpstra (1962)]. An approximate calculation based on an experiment where glucose was added to washed bacteria [Gitel'zon and Chumakova (1965); Gitel'zon *et al.* (1965*b*)] shows that bioluminescence accounts for only about a millionth of the total energy liberated in the oxidation of glucose. It was also estimated that a bacterial cell emits on average 100 quanta per second. Since the spectral distribution of one-electron noise of the photomultipliers (see Section 5.4) corresponds to the distribution spectrum of the signal from bacterial emission at the photomultiplier output, Fish *et al.* (1965*a*) propose a one-quantum character for bacterial luminescence.

Detailed description of the process underlying bacterial luminescence may be found in a number of reviews [Reid (1957); McElroy (1961); Chase (1964); Cormier and Totter (1964)]. We shall only note that the luminescence depends strongly on the pH of the medium, concentration of salts, inhibitors, and activators [Strehler and Cormier (1954)], concentration of oxygen [Gitel'zon *et al.* (1965*c*)], and also factors such as temperature and pressure [Lippincott *et al.* (1964)].

The connection between bacterial bioluminescence and the main metabolic reactions allows us to obtain various kinds of information about bacterial metabolism [see the reviews by Giese (1957); Chase (1964)]. Luminescent bacteria also proved to be a convenient model for studying the action of ionizing radiation [Burns *et al.* (1962, 1965); Bruckner (1964); Burns and Whipple (1964)]. Mention should be made of the work done on nonluminescent mutants of the luminescent organism, ***Photobacterium fischeri***. Using the aldehyde-free mutant (needing added aldehyde for luminescence), Rogers and McElroy (1958*a*, 1958*b*) studied the penetration of aldehyde into bacterial cells, and showed that (1) there is no special mechanism responsible for the transport of aldehyde and (2) the lipid solubility is of little importance: the determining factors are the size of the molecule and of the pores of the membrane.

Bioluminescent systems of bacteria may serve as useful models for the study of some processes of protein synthesis. In many luminescent systems the luciferase catalyzing the light-producing reaction has been shown to be a protein. The detection of luminescence may become a convenient way to follow the biosynthesis of luciferase.

The interest in luciferase synthesis seems to us to be particularly justified in that luciferase is a protein having a prosthetic group and consists of several subunits, and there are still many "blanks" in the study of the synthesis of such proteins.

In this connection mention should be made of the work of Karakashian and Hastings (1962, 1963), who studied the action of macromolecular synthesis inhibitors on the rhythm of the bioluminescence of the dinoflagellate, *Gonyaulax*. These organisms have one of the elementary systems of bioluminescence (Table 4.2). Inhibitors of biosynthesis actinomycin D and puromycin

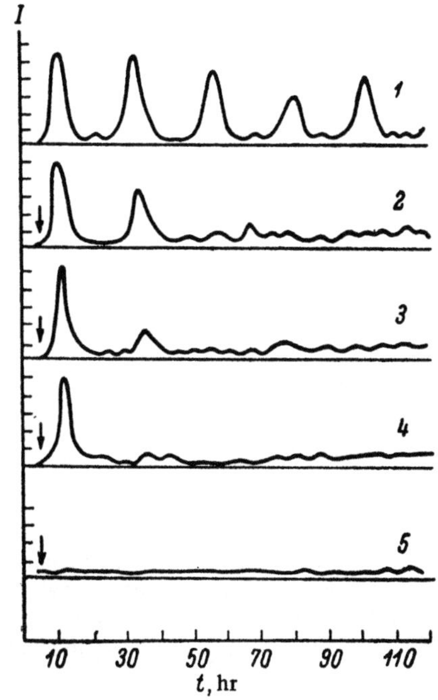

Fig. 4.5. Action of actinomycin D and puromycin on the rhythm of the luminescence of *Gonyaulax* [after Karakashian and Hastings (1962)]. (1) Control; (2) actinomycin D, 0.02 mg/ml; (3) actinomycin D, 0.04 mg/ml; (4) actinomycin D, 0.08 mg/ml; (5) puromycin, 10^{-5} *M*. (The arrows indicate the moment of addition of the antibiotic.)

disturbed the rhythm of the bioluminescence (Fig. 4.5), a fact which was connected with suppression of RNA synthesis. In our opinion, a system of luminescent bacteria may serve as a model for studying protein synthesis if the bacterial luciferase is synthesized according to the general scheme DNA → RNA → protein. An advantage of the use of bacteria is one that can obtain luciferase-free mutants [Rogers and McElroy (1955)] growing as well as the wild type.

An additional complicating factor in the utilization of bacteria for such investigations is the fact that bacterial glow is associated with respiration, which is also affected by protein-synthesis inhibitors. In view of this I. P. Novikov, of the authors' laboratory, constructed for this purpose an apparatus based on Warburg's apparatus, the six containers of which were connected with photomultipliers coupled to an automatic recording system. For another model Novikov studied the possibility of simultaneous automatic recording of the glow intensity, the growth rate, and (with the aid of electrochemical transducers) respiration. Preliminary experiments on inhibition, carried out jointly with N. A. Belozerskaya, indicated that in principle luminescent bacteria can be used to study kinetic aspects of biosynthesis.

The Bioluminescence of Jellyfish

Considerable attention has recently been directed to the luminescence of *Aequorea* and *Halistaura*, which for this purpose require only a protein, calcium

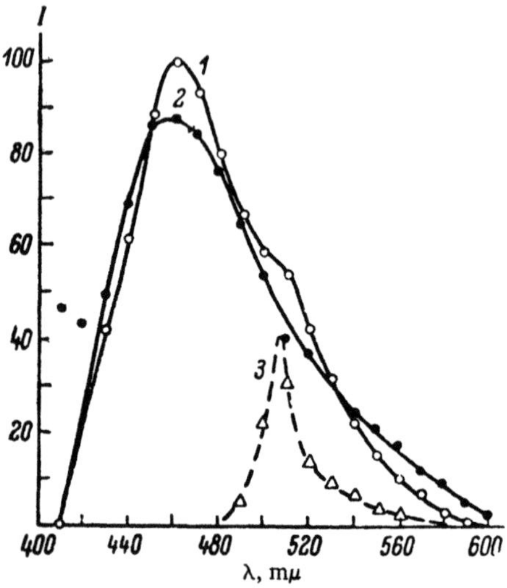

Fig.4.6. Fluorescence spectra of the material, aequorin, and of "green protein," obtained from *Aequorea* [Johnson *et al.* (1962)]. (1) Material extracted from photogenic tissue after storage for several days in a saturated solution of ammonium sulfate; (2) purified solution of aequorin extracted from photogenic tissue; (3) "green protein" (λ_{exc} = 390 or 460 mμ) extracted from photogenic tissue.

or strontium ions, and water. In contrast to the overwhelming majority of known systems, the presence of oxygen is not required for the luminescence of jellyfish. A proteinaceous material expressed from the photogenic tissues of *Aequorea* had a blue-green luminescence with a maximum at 460 mμ and a small "shoulder" at 500–520 mμ. Two substances participating in the luminescence were then isolated from this material [Johnson *et al.* (1962); Shimomura *et al.* (1962*a*, 1962*b*, 1963*a*, 1963*b*)]. The first turned out to be a protein (aequorin) characterized by a molecular weight of 3.5×10^3, a sedimentation constant of 2.1 ± 0.5 S, an isoelectric point at pH < 5.5, and an absorption spectrum having a maximum at 280 mμ and a "shoulder" at 310 mμ. Addition of Ca^{2+} to a solution of aequorin results in glow with a maximum at 465 mμ. The second substance ("green protein") has not yet been investigated. On excitation in the region of 270 mμ this protein fluoresces with two peaks, at 350 and 510 mμ, and on excitation in the region of 390 or 460 mμ it shows a maximum at 515 mμ.

Such overlapping of the emission spectrum of aequorin and the absorption spectrum of green protein (Fig. 4.6) ensures the possibility of the migration of excitation energy from one protein to another (or of trivial reabsorption of light).

A protein resembling aequorin in its physicochemical properties was isolated from the photogenic tissues of *Halistaura* and called halistaurin by analogy with aequorin [Shimomura *et al.* (1963*b*)]. The absorption spectra of halistaurin and aequorin, having maxima at 280 and 310 mμ, indicate that both these substances contain both protein and nonprotein components. The nonprotein component has in fact been discovered in the reaction products by spectral characteristics, and has been found to have absorption and fluorescence properties similar to those of reduced $NADH_2$ (DPNH). The authors suggest that this pyridine component is the source of light emission.

The basic scheme of the light-producing reaction is thus production of excited aequorin in the presence of Ca^{2+}, and the pyridine component of aequorin emits the light. Part of the excitation energy migrates to the green protein (or reabsorption of light takes place), resulting in the blue-green glow of the jellyfish. The physical mechanism of the excitation of aequorin or of the formation of its excited complex with Ca^{2+} is still unexplained.

The protein of aequorin obtained from the jellyfish *Aequorea* allows a rapid and fairly simple determination of small amounts of Ca and Sr ions in biological systems [Shimomura *et al.* (1963*c*)].

The Biological Role of Bioluminescence

In certain multicellular organisms luminescence has the function of attracting individuals of opposite sex, enticing prey [Harvey (1952); Reid (1957)], and protection from enemies [Clarke (1963)], which might have promoted survival of species in the course of evolution. This point of view is not new: it can be found in papers by Sizov (1917), Grinberg (1920), Shimkevich (1923), and perhaps in still earlier works. J. B. Buck (1961) proposed that in early stages of evolution there was a connection between visual and bioluminescent organs. In most light-emitting unicellular organisms the luminescence evidently has no adaptive significance, and is often regarded as a side effect of metabolic activity. However, Chumakova (1963*b*) believes that luminescence in bacteria is a form of protection of the cell structure from energy that for some reason had not been utilized. And indeed, if bacterial bioluminescence serves no useful purpose it is not clear what stops the dark mutants from replacing the luminescent forms.

The fact that the chemical basis of bioluminescence is different in different organisms agrees with the lack of any regularity in the distribution of luminescent species over the large taxonomical categories. McElroy and Seliger (1962*b*, 1962*d*) proposed that on the evolutionary path from anaerobic to aerobic life oxygen was toxic for the anaerobes, which consequently developed a method of dealing with any entering molecular oxygen by using it up on oxidation of luciferin. Thus, in their opinion, bioluminescence is a rudimentary phenomenon of adaptive significance at early stages of evolution. This point of view has been examined critically by Zhuravlev (1965*a*, 1965*b*). According to Zhuravlev's original hypothesis, bioluminescence appeared as the primordial internal mechanism supplying energy for the needs of the organism. With the development of the chain of oxidation–reduction enzymes the need for such a source disappeared and bioluminescence began to protect the biostructures in tissues of higher organisms against spontaneous oxidation. Moreover, in the evolutionary process certain organisms began to utilize bioluminescence for purposes of information or deception.

The evidence is as yet too inconclusive to allow a definite decision about the role of bioluminescence. Its molecular mechanisms have only relatively recently been subjected to intent study. And it is possible that it is in the course of deciphering these mechanisms that we shall discover the biological role of bioluminescence in organisms in which today bioluminescence is regarded as devoid of practical significance.

4.3. INDUCED CHEMILUMINESCENCE

In certain cases chemiluminescence of living objects and biologically important substances appears when they are subjected to the action of light, emission from radioactive sources, X rays, or finally heat. The products formed as a result of such influences enter into various chemical interactions between themselves, with the starting substance, and with atmospheric oxygen. Some of these reactions are accompanied by emission of light, and the study of this emission is important because it leads to an understanding of the interaction between the inducing radiation and the substance. Nevertheless, we shall not go into the finer details of such chemiluminescence, because clarification of these problems would require preliminary exposition of a fairly wide background in various fields of photochemistry, radiation chemistry, and thermochemistry, in which the authors have not specialized. Instead, we shall restrict ourselves to a phenomenological description, indicating the appropriate bibliographic sources.

Photochemiluminescence

Photoactive Objects. Strehler (1951) and Strehler and Arnold (1951) described a prolonged afterglow of photosynthesizing organisms in a suspension of chloroplasts. A relatively large amount of work was devoted to the nature of this phenomenon, some of which is analyzed in papers by Brugger and Franck (1958) and by Litvin *et al.* (1960). The induced luminescence of chlorophyll, its analogs, and chlorophyll-containing cells is thought to be due to oxidation of photoreduced pigments by oxygen and oxidation of pigments by peroxide [Krasnovskii (1963); Vladimirov and Litvin (1965)].

Nonphotoactive Objects. Vladimirov and Litvin (1959) discovered a prolonged emission from dry proteins irradiated with ultraviolet rays at room temperature. This emission was subsequently studied in greater detail by Konev and Katibnikov (1961*b*). In 1962 the latter two authors discovered in this emission two components, with long (hours) and short (seconds) lifetimes, and proposed that the afterglow is due to conservation of excited electrons at a supermolecular level of organization of the protein and subsequent emission from the triplet level of tryptophan. Analysis of the results obtained by Konev and Katibnikov led to the conclusion that this luminescence too is the result of certain photochemical reactions [Barenboim (1963*b*)].

Sapezhinskii *et al.* (1964*a*, 1965*a*), Sapezhinskii and Silaev (1965*a*, 1965*b*), and Sapezhinskii and Emanuel' (1965) showed that the long-lived component has a chemiluminescent character and is due either to interaction of peroxides with reactive groups in the protein [Sapezhinskii and Emanuel' (1965)] or, more probably, to recombination of peroxide radicals formed by irradiation [Sapezhinskii *et al.* 1965*a*)]. The nature of these radicals and the actual mechanism of their interaction are still unknown.

Living tissues which do not normally luminesce are also sometimes susceptible to photo-induced chemiluminescence. Thus, Vladimirov *et al.* (1962) observed this phenomenon in yeast; the emission maxima were at 460 and 510 mμ. Fish *et al.* (1965*b*) reported an enhancement of the ultraweak emission from the surface of viable rat liver after UV irradiation. The mechanism of the induced emission of nonphotoactive cells has not been studied in detail.

Chemiluminescence Appearing in the Presence of Ionizing Radiation

The existence of a radical mechanism of the photochemiluminescence of proteins suggests that even stronger chemiluminescence will be obtained upon irradiation with X or γ radiation, which give rise to much larger energy yields of radicals than ultraviolet light. We have already mentioned the studies in which such an emission was observed, and shall limit ourselves here to a very brief statement of the situation.

X irradiation of DNA in solution leads to the appearance of chemiluminescence which is ascribed to the formation of thymine hydroperoxides [Zhizhina *et al.* (1965); Emanuel' *et al.* (1965)].

Sapezhinskii *et al.* (1963, 1964*b*, 1965*b*) and Sapezhinskii *et al.* (1964*a*) showed that recombination of peroxide radicals of proteins (arising as a result of ionizing radiation) leads to the chemiluminescence of various protein solutions. In dry irradiated proteins such recombination occurs as the proteins are dissolved, and is also accompanied by emission of light [Sapezhinskii *et al.* (1963)]. An analogous phenomenon was observed on immersing in water dry cells previously exposed to X rays [E.M. Brumberg and I. Ya. Barskii (1960)]. In this case the emission may be due not only to recombination of protein peroxide radicals but also to interaction of the radicals formed from other cytological substances. In this regard, emission of light during dissolution has been found in previously irradiated sugar, starch, and glucose [Westermark and Grapendiesser (1960)].

Chemiluminescence Appearing During Heating

Heating of various substances may lead to the formation of products of thermal decomposition, and reactions of these products between themselves and with oxygen are accompanied by emission of light (thermochemical emission). Heating may also promote the interaction of products formed previously by irradiation of the parent substance with light, X rays, γ, α, or β radiation, and so on, especially if such interactions had been arrested by low temperature. This process will be referred to as thermoluminescence.

Thermochemical Emission. Mochalkin (1965) found that heating of almost all investigated powdered amino acids (alanine, asparagine, valine, leucine, tryptophan, tyrosine, etc.) leads to the appearance of an emission in the region of 380–540 mμ. The emission appears at temperatures of 120–150°C or higher, but only in the presence of air or oxygen; no emission was detected in helium or argon.

Assuming that an oxidation reaction was responsible for the emission, Mochalkin determined the activation energy ΔE of this process for various amino acids. The similarity of the ΔE values for a number of amino acids (0.65–0.70 eV) indicates that the mechanisms of the oxidative decomposition are similar, though some exceptions have been found; thus phenylalanine has two activation barriers, corresponding to two emission maxima.

The specific mechanism is still unclear. The values of ΔE probably refer to various reactions, and it may be that such averaging results in the similarity of the 'overall' activation energies. Moreover, we can hardly accept the author's hypothesis that the experimentally determined values of ΔE are similar to those in the corresponding processes in living tissue; there is no need for any proof that the oxidation processes occurring during the thermal destruction of biologically important substances and the latter's oxidation in the organism may proceed by completely different routes.

Induced Thermoluminescence. Depending on the mechanism of the reaction leading to the emission, this phenomenon may be divided into thermoluminescence and thermochemiluminescence.

In its classical form thermoluminescence appears when electrons escape from traps as a result of heating (after the traps become filled at low temperatures) and then recombine with positively-charged centers. The traps may be filled by irradiating the substance with light or ionizing radiations at a sufficiently low temperature. When the temperature is raised at a rate β, the substance begins to glow with an intensity reaching its maximum at a temperature T^*. The values of β and T^*, half-widths δ of the glow peaks, and certain other characteristics allow the determination of the energy depth of the traps E, the energy distribution $N(E)$ of the number of traps N, and the distribution of electrons over the traps $n(E)$. This method of investigation was proposed by Urbach (1930) and was later developed by many other authors. A fundamental review of the associated theoretical work has been published by Lushchik (1955).

Thermoluminescence is thus a physical phenomenon having a fully defined nature, and is not due to chemical destruction of the molecule. On the other hand, not every case of emission appearing during the heating of preirradiated objects can be called thermoluminescence, although this is often done. Thermoluminescence must be distinguished from thermochemiluminescence, i.e., the emission appearing in the course of heat-activated chemical reactions between the products of photolysis or radiolysis (possibly with the participation of oxygen).

Thus, for example, Arnold and Sherwood (1957) discovered an emission appearing on heating of previously irradiated chloroplast membranes from various plants, which they associated with the purely physical process of energy storage in semiconductor electron traps. The phenomenon was later studied by Litvin *et al.* (1960) and Tumerman *et al.* (1962), who showed that it can be explained by the formation of a peroxide in the course of illumination; this peroxide then decomposes during the heating, i.e., the emission is thermochemiluminescence.

Thermoluminescence is evidently a general phenomenon of a different kind than chemiluminescence, but in many practical cases it is very difficult to say just what is the nature of the thermally induced luminescence (thermoluminescence or thermochemiluminescence). It therefore seems useful to give at this point the results of studies in which the authors believed that they were dealing with thermoluminescence or thermochemiluminescence.

Up to the present time investigations have been carried out on the induced thermoluminescence of deep-cooled (77°K) crystalline powders of amino acids and proteins previously irradiated with ultraviolet light [Gill (1963); Gill and Weissbluth (1964); Kuznetsov and Kukushkin (1965)], γ rays [Augenstein *et al.* (1960, 1961); Weinberg *et al.* (1962)], and X rays [Prydz and Rogeberg (1964)], and also of UV-irradiated frozen solutions of these compounds [Roshchupkin and Vladimirov (1964, 1965*a*, 1965*b*); Vladimirov *et al.* (1965)].

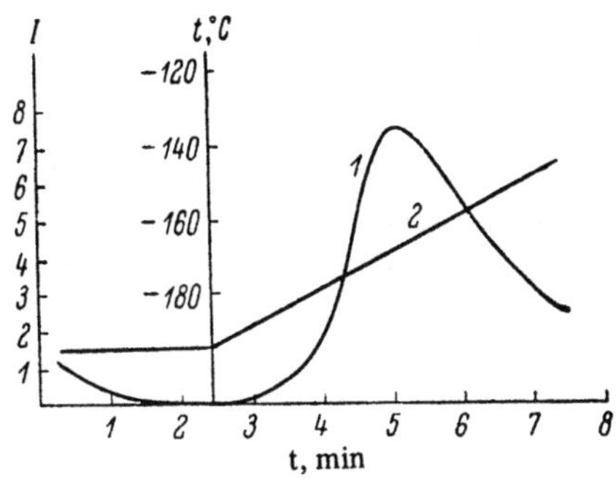

Fig.4.7. Thermoluminescence from tryptophan. (1) Thermoluminescence intensity (*I*) and (2) temperature *t*°C.

It is not difficult to detect the thermoluminescence from frozen solutions or dry powders of aromatic amino acids – the intensity of the emission from frozen solutions of tryptophan previously irradiated with ultraviolet is fairly high. As an example, Fig. 4.7 shows the thermoluminescence of crystalline tryptophan powder (unpublished work of the present authors together with V. F. Solomatin).

Induced thermoluminescence of biologically important substances was first investigated by Augenstein and his co-workers [Augenstein *et al.* (1960, 1961); Weinberg *et al.* (1962); Carter *et al.* (1963)] on powdered crystalline amino acids and dried proteins irradiated at liquid nitrogen temperatures with the γ rays from Co^{60}. The emission was ascribed to the formation of traps on irradiation, i.e., the emission was regarded as thermoluminescence.

According to the above authors the appearance of traps responsible for the thermoluminescence is due to the fact that anisotropy of the charge distribution in the protein leads to the formation of numerous potential wells capable of capturing holes as well. In a crystalline protein the number of traps increases owing to lattice defects. Semiconductor traps in dry proteins are also postulated in several other works [see the review by Eley (1962)]. However, the existence of semiconductor properties in proteins has also been criticized [Vol'kenshtein (1964)]. It is not impossible that the thermally induced emission or some part of it is due to the recombination of radicals appearing during the irradiation of dry cooled protein. In this context it is interesting that the activation energy for trypsin emission is 2 kcal/mole [Augenstein *et al.* (1961)] and the activation energy of the radical chemiluminescence reaction of previously irradiated serum albumin 2 – 10 kcal/mole [Sapezhinskii *et al.* 1965*a*)].

In the above studies Augenstein and his co-workers established that thermoluminescence is determined not so much by the structure of the crystal lattice as by the chemical structure of the molecules. This conclusion follows from the considerable differences in the intensity and the position of the temperature peaks of the emission curves for crystals of a number of amino acids which have the same crystal structure.

It was also found that the thermoluminescence of proteins is not the simple sum of the contributions due to independent amino acids making up these proteins. This demonstrates the effect exerted by protein structure on the thermoluminescence of amino acid residues. Evidently this conclusion applies so far only to the primary structure: the mere inclusion of amino acids into a polypeptide chain may change sharply their thermoluminescent properties by the formation of peptide linkages between α-amino and carboxyl groups. It would therefore be very interesting to investigate the thermoluminescence of simple dipeptides.

In 1962 Gill [cited in Gill and Weissbluth (1964)] observed thermoluminescence after preliminary UV irradiation of deep-cooled crystalline preparations of amino acids and proteins. The above authors regarded the phenomenon as thermoluminescence. The nature of the metastable level (trap) responsible for the storage of energy is unknown, but it is supposed that a possible mechanism by which this trap could be formed is capture of an electron by an NH_3^+ group (the excited group or one adjacent to it) or by metal ions (Fe^{3+}, Cu^{2+}, etc.) which are present in specimens of amino acids.

It is also supposed that the emission occurs during transition from a triplet level to the ground state [maximum in the region of 400–450 mμ Gill (1963)], which is preceded by an activating transition from the metastable level (trap) to the triplet level. For example, the activation energy for tyrosine was found to be 0.16 eV. The luminescence process is also activated by visible light: after prolonged illumination the thermoluminescence intensity falls by a factor of 20. The authors indicate that the thermoluminescence of proteins does not correspond to the monomolecular mechanism valid for the amino acids.

If we accept the possibility that a metastable level may be formed by electron capture by a group NH_3^+, it becomes understandable that the inclusion of amino acids into a polypeptide chain should lead to differences between the thermoluminescence of free amino acids and proteins.

Thermoluminescence of frozen solutions of amino acids and proteins irradiated with UV light was studied in detail by Roshchupkin and Vladimirov (1964, 1965*a*, 1965*b*) and by Vladimirov *et al.* (1965); see also the review by Vladimirov (1965*a*). The temperature range for maximum emission of all the investigated compounds was at 81–84°K.* Comparison of the thermoluminescence curves with the curves of slowly decaying emission first described by Debye and Edwards (1952) led to the conclusion that both these processes, spectrally close to phosphorescence, have a common nature and are due to the same unstable primary photoproduct P_t.

The above authors (Vladimirov and co-workers) believe that the product responsible for the thermoluminescence appears after a preliminary transition of a molecule A into an excited triplet state A_T*

$$A + h\nu \longrightarrow A_S^* \longrightarrow A_T^* \longrightarrow P_t,$$

and the thermoluminescence is due to the reverse photochemical reaction

$$P_t \longrightarrow A_T^* \longrightarrow A + h\nu_{phosphi}$$

*This is not in accordance with earlier results [Gill (1963)], according to which the thermoluminescence peak of, e.g., tyrosine appears at 124°K, and the curves of phenylalanine have two peaks, at 113 and 155°K, which are due to the existence of two metastable levels (traps) of different energies. These discrepancies may be ascribed to different starting states of the materials used in the above studies and to the fact that different rates of heating may affect sharply the position of a given peak and its resolution on the corresponding curves [Lushchik (1955)].

These mechanisms are discussed in detail by Vladimirov and Litvin (1965) and by Vladimirov (1965*a*).

As was already reported earlier [Gill and Weissbluth (1964)], in the case of powders thermoluminescence of frozen solutions falls sharply when a frozen specimen already preirradiated by UV light is then also illuminated by visible light. Roshchupkin and Vladimirov (1965*a*, 1965*b*) ascribe this phenomenon to the high light sensitivity of the photoproduct P_t, while Gill (1963) believes it to be due to the emptying of traps.

Douzou *et al.* (1961) tried unsuccessfully to detect thermoluminescence from DNA on UV irradiation.

In surveying these investigations on induced thermoluminescence, it is easy to see that in many of these studies the external picture of the phenomenon (spectral composition of the emission, the dependence of thermoluminescence on temperature, presence of gases, and additional illumination with visible light, the magnitude of the activation energy, and the differences between the thermoluminescence of free amino acids and proteins) is very much the same. The observed differences between these characteristics may be accounted for by differences in the starting states of the investigated materials (crystalline or dry substances, frozen solutions), by differences in the action of the radiation used to induce the thermoluminescence (UV light, X rays, γ rays), and finally by different experimental conditions (duration of irradiation, rate of heating, etc.).* Nevertheless, the nature of the phenomenon is explained in some works on the basis of the concept of semiconductor thermoluminescence and in other works on the basis of thermochemiluminescence due to radical mechanisms.

Can we in fact assume that factors such as the state of the substance and the type of the inducing radiation are responsible for fundamentally different mechanisms, or do they have only a slight effect on the phenomenological aspects? Is it possible both mechanisms are involved, either one or the other predominating in dependence on these factors?

The answers may be found by further work in which induced thermoluminescence will be studied varying the above factors under otherwise identical experimental conditions.

---

*The fact that thermally induced emission characteristics depend strongly on the type and energy of the ionizing radiation used in the irradiation stage follows from the work of Prydz and Rogeberg (1964), who determined the thermoluminescence spectra of tyrosine and tryptophan powders arising after irradiation. Such spectra are very different from the usual fluorescence and phosphorescence spectra of these amino acids (see Section 1.1), and also from thermoluminescence spectra appearing on γ irradiation. For tryptophan the spectral maxima were observed at 575 and 630 mμ, and for tyrosine at 490 mμ.

Chapter 5

APPARATUS FOR BIOLUMINESCENCE MEASUREMENTS

The principles underlying luminescence measurements can be summarized as follows. Radiation from the source is resolved into a spectrum by a monochromator or a set of light filters and is then intercepted by the object under investigation. The luminescence excited in the object is analyzed by a set of light filters, or is resolved into a spectrum by a monochromator, and is then intercepted by a radiation detector, the signal from which passes to the recorder.

All the necessary apparatus is available commercially and the problem for the experimenter is to select the appropriate components and use them to their best advantage. This is particularly important in the case of radiation detectors because improvements in detection techniques are the principal means of increasing the sensitivity of the apparatus as a whole.

Methods of resolving radiation into a spectrum have remained essentially the same since the work of Wood. Sources of UV radiation have been available for a considerable time and their properties are well known. However, it is only since the advent of the photomultiplier (PM) that detailed studies of the UV fluorescence of proteins and nucleic acids have become possible. The development of photomultipliers and, in particular, the development of methods for reducing the sensitivity threshold of these devices has resulted in the discovery of dark chemiluminescence.

In this chapter we shall give a brief review of the basic components of apparatus used in studying the luminescence of biopolymers and cells. Particular attention will be paid to applications of photomultipliers to the detection of very weak light fluxes.

5.1. CHOICE OF APPARATUS

In this section we shall discuss the light sources, monochromators, light filters and photomultipliers used in luminescence studies.

Light Sources

Table 5.1 gives a summary of the parameters of various UV sources used with biological specimens. They are gas-filled tubes in which the radiation is emitted as a result of an arc discharge. Their spectral characteristics and the associated circuits have been described in detail in the literature [Golland (1961); Konstantinova–Shlezinger (1961); Shishlovskii (1961); Vladimirov and Litvin (1964)]. Agroskin (1957) compared the luminance of a number of these tubes (SVD-120A, DRSh-100, DRSh-250, etc.) in the range 248–436 mμ with the PRK-4 tube.

The SVD-120A tube is particularly useful. It can be operated under AC or DC conditions, it is self-triggered, it emits copious radiation in the 240–280 mμ region, and can be used in conjunction with the commercially available condenser OI-18. This tube has therefore been widely used in the luminescence analysis of biological material. The DKsSh-200 tube is particularly useful for the recording of excitation spectra when the source of radiation must have a continuous spectrum in the region in which the object under investigation absorbs.

Arc-discharge sources have an unstable light output. This is due to variations in the supply voltage, displacement (including rotation) of the light column in the tube, and variations of the transmission properties of the envelope

Table 5.1. Characteristics of Tubes Used in the Luminescence Analysis of Biological Objects

					Operating conditions		
Type of tube	Spectrum in the UV region	Light flux, lumens	Intensity at center of discharge, minimum, Mnit	Distance between electrodes, mm	Supply voltage, minimum V	Voltage across tube, maximum, V	Working current through tube, maximum, A
GSVD-120	Continuous	–	40	3.0–3.5	150	25	5.0
DKsSh-200	"	–	70	2.0–2.5	70	22	12.0
DKsSh-1000	"	28000	200	2.7–4.1	70	20	50.0
DRSh-100	Line	1200	700	0.5	127	26	4.0
SVD-120A	"	4200	–	30	220	140	1.7
DRSh-250	"	12500	100	3.9	127	72	4.0
DRSh-500	"	22500	130	4.6	220	75	7.5
DRSh-1000	"	53000	120	8	220	90	12.0

The number following the tube designation indicates the power consumption in W. The first three tubes are DC operated; the remainder are DC or AC operated. The tube characteristics are based on the data supplied by the Moscow Electron Tube Factory.

with time. Stabilization of the supply voltage for such tubes does not suppress the other sources of instability, and therefore stable light outputs are usually produced by means of circuits which react directly to the emitted radiation. Such stabilizers have been described by Shirokov (1956) and Rusinov (1962) for the DRSh-250 tube and by Barenboim (1962) for the SVD-120A tube. The stabilization method put forward by Chance (1940) employs a photocell or photomultiplier to record the light output as a function of time. The resulting signal is then amplified and is used to control tubes shunting the light source. A block diagram of the system used in our own work is shown in Fig. 5.1. Radiation emitted by lamp (1) and transmitted by filters (2) is partly reflected (approximately 4%) by the quartz plate (3) onto the photocell (4). The photocell output is amplified by a two-stage DC amplifier and is applied to the grid of a 6N5S tube (Fig. 5.2). This tube shunts the SVD-120 light source, so that the anode voltage for both of them is derived for a single rectifier set. This system is capable of stabilizing the light output in a narrow spectral region to within ± 0.5% for a supply-voltage variation of ±30 V. This system suffers from the disadvantage that it does not react to rapid displacements of the light column in the source. Other methods of stabilization of the light output were reported by Mekshenkov and Andreitsev (1960) and by Eisenberg *et al.* (1962). Pulsed light sources have been used in luminescence studies of objects susceptible to rapid photochemical decomposition, in phosphorescence studies using pulsed excitation, and so on. I. Ya. Barskii and Yakubenas (1964) used the ISSh-500 tube in pulsed microphotography of biological objects. We have used the ISK-25 tube for phosphorescence excitation. The high-pressure tubes DRSh-250 and DRSh-200 can also be used successfully under pulsed conditions. In this case a 2000-μF capacitor charged to 200 V is discharged through the tube when the high voltage pulse is applied to the trigger electrode. Results show that these tubes have certain advantages under pulsed conditions as compared with the ISK-25 tube, i.e., their light output is more readily focused on the object, their spectrum is more

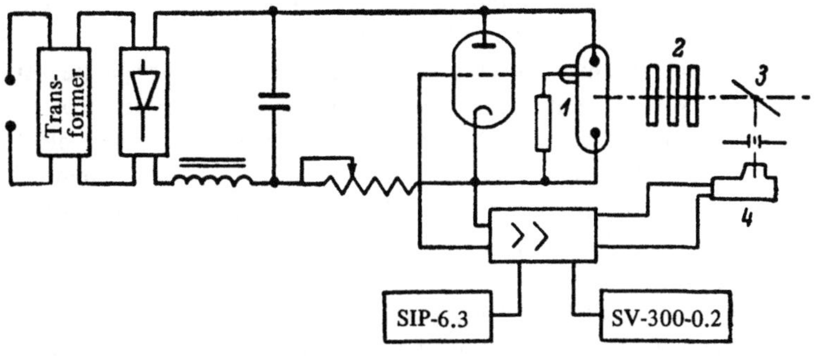

Fig. 5.1. Block diagram of the tube stabilizer (see text for explanation).

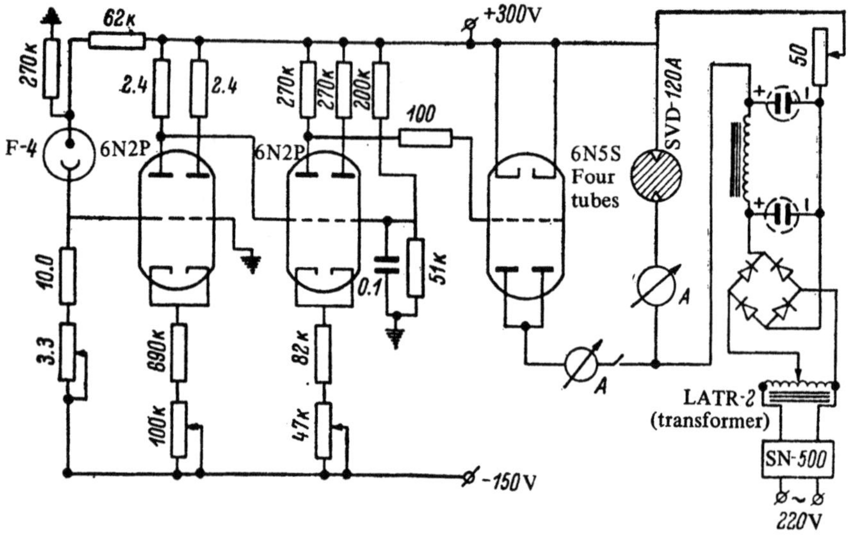

Fig. 5.2. Basic circuit of the stabilizer.

suitable for the excitation of bioluminescence because it is rich in the ultraviolet, and in contrast to the ISK-25 tube, the proportion of red radiation is lower. Unfortunately the power per pulse cannot be increased by varying the applied voltage because the breakdown voltage in these tubes is relatively low (between 200 and 300 V for the DRSh-250 tube and a little lower for the DRSh-500). Soviet pulsed tubes were reviewed by Marshak (1962, 1963).

Monochromators

It is desirable to use double monochromators for the excitation of the luminescence of biopolymers and cells. This eliminates scattered light in the spectrum which may mask the weak radiation emitted by the object. The DMR-4 monochromator is suitable for this purpose although the earlier version of this instrument, namely the DMR-2, is better.

High-focal-ratio monochromators are used to resolve the spectra, especially low-intensity spectra, but steps have to be taken to reduce scattered and absorbed radiation. Commercially available monochromators suitable for this purpose include the ZMR-3 monochromator and the monochromators in the SFD-2 and SF-4 spectrophotometers. These monochromators can also be successfully used to excite luminescence [Katibnikov and Konev (1961); Burshtein and Vladimirov (1964)].

Vasil'ev *et al.* (1963) developed a special monochromator which is of particular interest. This is based on the Fastie system and incorporates a replica rating (150 x 140 mm; 600 lines per mm). It has a focal ratio of 1:1.5, slit height of 70 mm, and is designed for the spectral analysis of ultraweak radiation. The MDR-2 monochromator is optically similar to that described above and has been discussed by Aleksandrov *et al.* (1965). Table 5.2 summarizes some of the parameters of commercially available UV monochromators.

Monochromator design was discussed by Shishlovskii (1961) and Toporets (1955). Commercial monochromators manufactured abroad and used for the luminescence analysis of biological objects were discussed by Udenfriend (1962). Monochromators for microspectroscopic studies were developed by Agroskin and Korolev (1963).

Light Filters

Light filters are widely used in the luminescence analysis of biopolymers and cells. They are used to define the necessary spectral region in the case of excitation, to define the luminescence spectrum and to reduce scattered radiation. Moreover, they are the only means of estimating the spectral composition of ultraweak light fluxes when high-focal-ratio monochromators become ineffective. Carter *et al.* (1964) described an instrument in which 50 narrow-band light filters are used to examine the spectrum of ultraweak ultraviolet and visible radiation. A method which can be used to determine the emission spectra with a set of broad-band light filters was described by Vasil'ev (1965).

Narrow-band interference light filters whose operation is based on the phenomenon of multiple-beam interference can be used to define spectral regions of the order of 10 Å with a transmission factor at maximum transmission

Table 5.2. Some Characteristics of Soviet UV Monochromators

Type	Working region, mμ	Focal ratio	Focal length of exit collimator, mm	Dispersing element	Dispersion D_l at wavelength λ: D_l, Å/mm	λ, mμ
DMR-4	210–2500	1:9.4	343	Two 30° prisms	9.9	219
DMR-2	220–2700	1:7	270	Two 56° prisms	6	220
ZMR-3	220–2500	1:7	270	One 56° prism	29	271
Monochromator from the SF-4 spectrophotometer	210–1100	1:10	500	One 30° prism	7.5 100	200 400
Monochromator from the SFD-2 spectrophotometer	210–1100	1:10	500	Plane replica grating (600 lines/mm)	32	Throughout working region
MDR-2	200–2500	1:2.4	400	Three demountable plane replica gratings (300, 600, and 1200 lines/mm)	Not given	

of about 50–60%. Such filters were described in detail by Golubeva *et al.* (1965). Interference light filters for the ultraviolet part of the spectrum are now available in the United States.

Absorption light filters are usually made of colored glass, an absorbing solution, or a gas. Glass filters have broad transmission bands and are available commercially. Suitable gases for the gas-filled filters include chlorine, bromine, and their mixtures. Liquid filters for the ultraviolet region are discussed by McLaren and Shugar (1964). These workers have given the compositon of cut-off filters and narrow-band filters for the 240–380 mμ region with a transmission band of about 30 mμ. Van Es and Wisse (1963) have described a number of narrow-band filters for the excitation of the luminescence of proteins. Some information about liquid filters can be found in the books by Meier and Zeitts (1952) and Shishlovskii (1961).

It is important to note that the transmission of some of the UV light filters varies with time as the result of the solarization in the case of glass filters, and decomposition in the case of liquid filters. Some filters luminesce under the action of short-wave UV radiation (especially the BS-type filters).

Figure 5.3 shows the transmission of a set of light filters usually employed in the observation of the luminescence of proteins and cells, together with the excitation and fluorescence spectra.

Photomultipliers

Photomultipliers are used in practically all types of apparatus for the detection of radiation emitted by biopolymers and cells. The physical phenom-

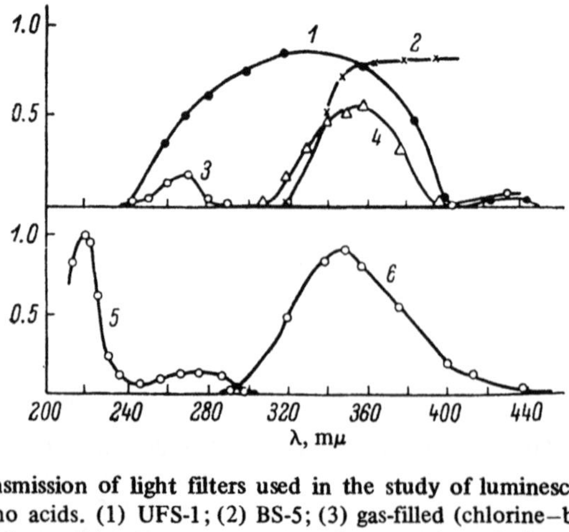

Fig. 5.3. Transmission of light filters used in the study of luminescence of proteins and aromatic amino acids. (1) UFS-1; (2) BS-5; (3) gas-filled (chlorine–bromine mixture); (4) UFS-6. Excitation and luminescence spectra of a protein solution are shown for comparison (curves 5 and 6).

ena which occur in these devices have been described in the extensive literature which is now available on the subject. This includes the monographs by Chechik *et al.* (1957) and Birks (1964).

It is important to note that in most books and experimental papers on photomultipliers these devices are discussed from the point of view of scintillation spectrometry where they are most widely used.

The chief parameters of a photomultiplier are: (1) spectral sensitivity, (2) integral sensitivity of the photocathode, (3) gain, and (4) dark current. Photomultipliers used for the detection of the luminescence of biological material must be sensitive in the ultraviolet and visible regions of the spectrum and also must be capable of recording light fluxes of the order of 10–100 photons per sec, but we are not aware of the existence of any photomultiplier having such "ideal" characteristics.

Table 5.3 gives a summary of the properties of Soviet photomultipliers capable of recording ultraviolet radiation. The characteristics of the photocathodes are given in Fig. 5.4.

When UV sensitive photomultipliers are not available, it is a common method to use radiation converters which luminesce under the action of ultraviolet radiation in the region in which the photomultiplier is sensitive. Such phosphors are used as proportional photon counters to produce the control signal necessary for the stabilization circuits of light sources and for the automatic correction of spectra (see Section 6.1).

A desirable property of a phosphor is that its luminescence energy yield is

Table 5.3. Parameters of Some Photomultipliers Sensitive in the UV Region

Type	Type of spectral sensitivity	Sensitive range	Position of sensitivity peak, mμ	Working voltage, V	Minimum integrated photocathode sensitivity, μ/lumen	Integrated area sensitivity, A/lumen	Maximum dark current, A	Maximum output current, mA
FEU-18A	S-3	215–600	380 ± 50	900	20	10	$3 \cdot 10^{-9}$	0.1
FEU-39	S-6	160–600	400 ± 20	1200	25	10	$3 \cdot 10^{-9}$	0.2
FEU-46A	S-9	200–650	390 ± 50	1000	30	2	10^{-10}	1.0
FEU-53	S-12	250–650	410 ± 10	1700	30	8	$4 \cdot 10^{-7}$	10
FEU-57	S-14	200–400	255 ± 25	1700			$5 \cdot 10^{-8}$	1

The table is based on nominal data supplied by the manufacturers. The type of spectral sensitivity is determined by the material used for the cathode, its base, and the window. The window transmission is the main factor in the UV and therefore the above photomultipliers have quartz or uviol windows. The analogous photomultipliers with glass envelopes (FEU-17A, FEU-37, and FEU-46) have spectral characteristics of type S-6 and are sensitive in the range 300–600 mμ.

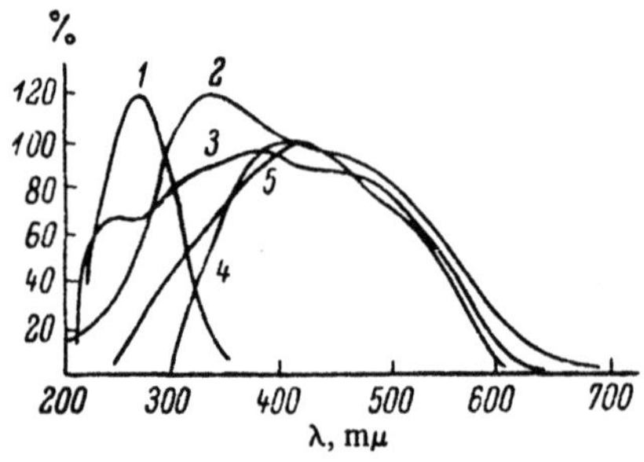

Fig. 5.4. Spectral characteristics of photocathodes. The relative sensitivity of the photocathode is plotted along the vertical axis: (1) S-24; (2) S-9; (3) S-3; (4) S-6; (5) S-12.

independent of the wavelength of the exciting radiation. The spectral sensitivity of the photomultiplier–phosphor system is then constant. In the 200–320 mμ region this property is exhibited by sodium salicylate films (λ_{max} ~ 405 mμ), esculin (λ_{max} ~ 490 mμ), anthracene (λ_{max} ~ 440 mμ), and so on. Uranium glass (ZhS-19) is frequently used as a phosphor (λ_{max} ~ 530 mμ). Khlebnikov *et al.* (1961) showed that a 0.1-mm uranium glass plate operating in conjunction with the FEU-46 photomultiplier will extend the spectral sensitivity range of the photomultiplier down to 140 mμ. The sensitivity in the 140–320 mμ region is practically constant. However, the use of a uranium glass plate as a phosphor suffers from certain disadvantages, namely, the emission maximum lies in the yellow-green region (λ_{max} ~ 530 mμ) which means that the sensitivity curve falls in the region where the SbCs photocathode is particularly sensitive (λ_{max} ~ 400 mμ). The luminescence excitation spectrum has a valley at 370mμ and the conversion efficiency is reduced when the glass plate is thicker than 0.5 mm. Meier and Zeitts (1952) and Chechik *et al.* (1957) summarized the characteristics of a number of phosphors capable of converting UV radiation. Soviet photomultipliers were described by Leiteizen and Glukhovskii (1964) (FEU-28, FEU-38, and FEU-51) and by Vil'dgrube *et al.* (1963, 1964) (venetian-blind types FEU-11, FEU-16, FEU-49, FEU-52–FEU-56). Photomultipliers manufactured abroad were described by Birks (1964). New types of photomultiplier are discussed by Matheson (1964) and by Stanley (1964). Photomultipliers with S-6 type photocathodes can be used for the detection of ultraviolet luminescence (excitation $\lambda_{max} < 300$ mμ; emission $\lambda_{max} > 300$ mμ). These photocathodes are insensitive to the exciting radiation (for example, the low-noise photomultiplier FEU-64).

5.2. METHODS OF REDUCING PHOTOMULTIPLIER NOISE

The dark current flowing through any photomultiplier, even when it is not illuminated, governs its sensitivity threshold and is due to the following factors: (1) thermoelectric emission by the photocathode and the dynodes, (2) leakage currents through the photomultiplier envelope, (3) field emission, (4) ion and optical feedback, and (5) radiation due to the K^{40} isotope in the glass envelope of the photomultiplier.

Various measures can be taken both in the design and the construction of the photomultiplier to reduce the dark current. For example, the thermoelectric component of the dark current is reduced by reducing the photocathode area (FEU-46) or by introducing special diaphragms between the photocathode and the first dynode. These diaphragms transmit photo- and thermoelectrons only from the central part of the cathode (FEU-64). Some photomultipliers are made from potassium-free glass (FEU-42) to reduce the photomultiplier noise due to K^{40} radiation falling on the photocathode and dynodes, and the noise due to scintillations in the glass envelope under the action of this radiation [Fleishman and Shakhidzhanyan (1959)]. This measure reduces the mean amplitude of the noise pulses.

There are many ways which can, in principle, be used to reduce photomultiplier noise. However, many of them are of limited value to the experimenter who has at his disposal only the commercially available photomultipliers; others do not ensure that the sensitivity threshold is at the required level. In such cases, there are a number of methods which can be used under laboratory conditions. These include selection of optimum operating conditions for the photomultipliers, magnetic defocusing lenses, electrostatic screens, and cooling to low temperatures. Each of these can be used either separately or in combination with others.

Suitable choice of dynode potentials will ensure that the photomultiplier is operated under maximum gain conditions [Chechik *et al.*(1957)]. This also improves the time resolution [Yakushin and Tel'nov (1963)]. The voltage between the photocathode and the first dynode is particularly important [Vil'dgrube and Ronkin (1962); Walton (1964)].

The application of a magnetic field to a photomultiplier [Naray (1956); Avakyan *et al.* (1963)] is equivalent to a reduction in the working area of the cathode. It reduces the noise level by roughly a factor of two. The use of this method is connected with the necessity of removing charges from the inner surface of the glass envelope, and requires good stabilization of the current in the magnetic lens.

An electrostatic screen (Aquadag layer, or metal cylinder around the photomultiplier tube) maintained at the photocathode potential will reduce the dark current in some photomultipliers by roughly two orders of magnitude [Naray (1956); Chechik *et al.* (1957); Hartmann and Bernhard (1957)]. When the screen potential is increased there is a rapid rise in the noise level which reaches a maximum when the screen is grounded. It is therefore important to prevent any possibility of contact between the photomultiplier envelope and grounded metal components, e.g., the light-tight envelope. It is important to note, however, that the use of a screen will not yield the desired effect in some photomultipliers. Pertsev *et al.* (1963) have found that an electrostatic screen maintained at the photocathode potential reduces the absolute number of noise pulses by an order of magnitude in the case of the FEU-A1 photomultiplier, but it *increases* this number in the case of FEU-42 and FEU-43. In our own laboratory A.M. Kudashev and A. Pluchennichak have shown that an electrostatic screen has no effect on noise in the FEU-37, FEU-39, and FEU-64, but reduces the noise level by about 40% in FEU-57.

The most radical method of reducing the noise level in mass-produced photomultipliers is cooling. The problem then arises as to what is the best method of cooling and what is the optimum temperature.

Comparison of Different Methods of Photomultiplier Cooling

The two main media which can be used for cooling photomultipliers are solid carbon dioxide (–78°C) and liquid nitrogen (–196°C). In the case of carbon dioxide [Vasil'ev *et al.* (1961)] the end-window multiplier is placed in a hermetically sealed steel envelope which in turn is placed in a plastic foam container. The space between them is filled with the carbon dioxide. The consumption of the cooling agent in this case is 0.5–1 kg/hr. The dark current in various photomultipliers is reduced in this way by 1–3 orders of magnitude (FEU-28, FEU-29, FEU-VEI-2V).

In the case of liquid-nitrogen cooling there are various methods including placing the photomultiplier in contact with a cooled metal plate [Chechik *et al.* (1957); Zakharov and Shlyapintokh (1961)], placing the photomultiplier in a liquid nitrogen container [Vladimirov and Litvin (1959*a*); Konev and Katibnikov (1961*a*); Tarusov *et al.* (1961*b*)], and placing the photomultiplier in liquid nitrogen vapor or in cooled nitrogen (Barenboim *et al.* (1964*b*); Franklin *et al.* (1965)]. The first method requires good contact between the plate and the photomultiplier envelope, and the second a dewar container of suitable size and an optical window facing the photocathode. Occasionally, the photomultiplier is inserted into a liquid-nitrogen cooled container [Osherovich *et al.* (1962); Gandy and Weller (1964)]. The temperature of the photomultiplier can then be regulated with a heater [Osherovich *et al.* (1962)]. By cooling the photomultiplier down to –196°C it is possible to reduce the dark current in some cases by four orders of magnitude. Vladimirov and Litvin (1959) have shown that in the case of the FEU-18 tube the noise is reduced down to 2–5 pulses/sec. A number of systems for photomultiplier cooling developed in our own laboratory are summarized below [Barenboim *et al.* (1964*b*, 1967*a*)].

Cooling of Photomultipliers by Liquid Nitrogen Vapor. The apparatus is illustrated schematically in Fig. 5.5. Liquid nitrogen is boiled off from a dewar by means of a heater, and the cold gas is introduced through a tube into a thermally insulated chamber containing the photomultiplier. By varying the rate of evaporation, the temperature inside the container can be varied between room temperature and –150°C. This method can be used to cool end-window photo-

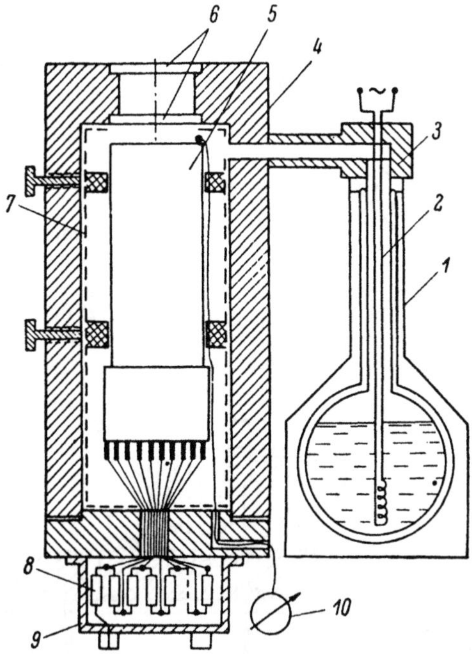

Fig. 5.5. Cooling of the photomultiplier by liquid-nitrogen vapor. (1) Dewar container (SV-15); (2) heater-evaporator; (3) vapor tube; (4) textolite envelope; (5) photomultiplier; (6) quartz windows; (7), (9) screen; (8) potential divider; (10). thermocouple with output microammeter.

multipliers with semitransparent photocathodes (FEU-39 and FEU-46) for which the optimum temperature depends on the intensity of the incident radiation [Barenboim *et al.* (1966)]. For the above photomultiplier types, and a constant rate of evaporation, temperature equilibrium is established in roughly one hour. It is possible to achieve temperature variations of less than ± 0.5°C at the photomultiplier window. The nitrogen consumption necessary to maintain temperatures of –150, –120, and –60°C is 15, 3, and 1 litres/h respectively. The intrinsic noise in FEU-39 and FEU-36A at –120°C is reduced roughly by a factor of 150 (Fig. 5.6).

Cooling of Photomultipliers Through the Photocathode and Dynode Pins. In some photomultipliers (FEU-17, FEU-20, FEU-22, etc.) the photocathode is supported inside the glass envelope by a massive metal piece. It is therefore sufficient to cool the metal leads into the photomultiplier. This method is illustrated in Fig. 5.7.

Figure 5.8 shows the photocathode temperature as a function of time of cooling, and the fall in the dark current in FEU-18. Temperature equilibrium is

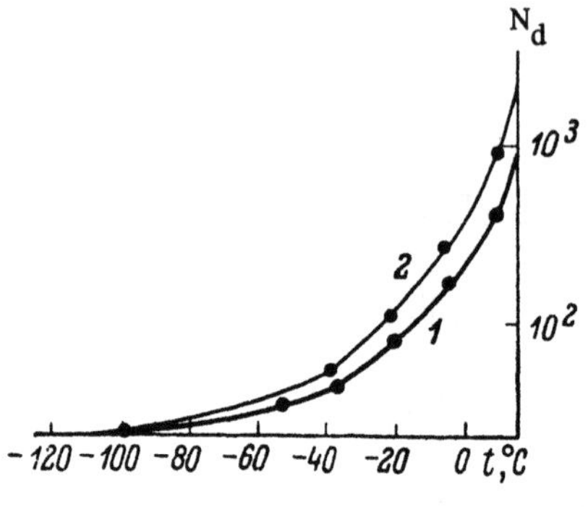

Fig. 5.6. Reduction in the photomultiplier dark current on cooling. Vertical axis shows number of pulses/sec (1) FEU-39; (2) FEU-46A.

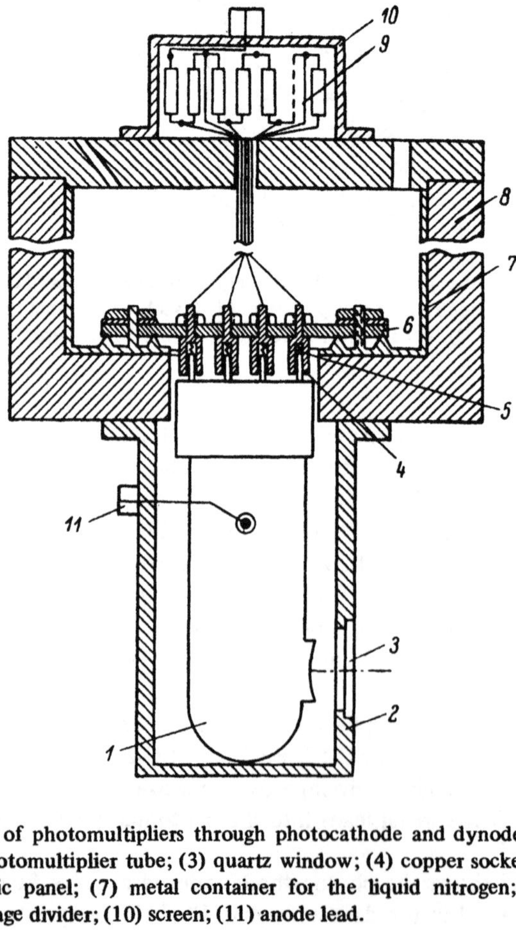

Fig. 5.7. Cooling of photomultipliers through photocathode and dynode pins. (1) Photomultiplier; (2) photomultiplier tube; (3) quartz window; (4) copper sockets; (5) contacting springs; (6) plastic panel; (7) metal container for the liquid nitrogen; (8) plastic foam insulator; (9) voltage divider; (10) screen; (11) anode lead.

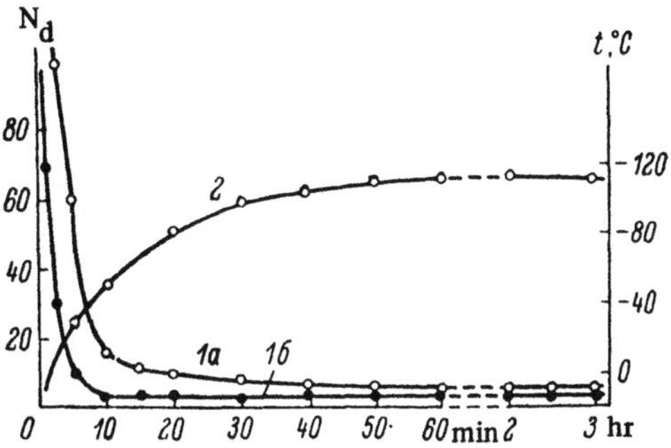

Fig. 5.8. Number of pulses per sec (left-hand axis) and temperature (right-hand axis) as functions of time of cooling for the FEU-18A photomultiplier. (1a) Intrinsic noise; (1b) photomultiplier immersed directly in liquid nitrogen; (2) photocathode temperature.

established in 50–60 min, at which time the temperature of the photocathode is –120°C and the dark current is 5–6 pulses/sec. Nitrogen consumption is about 4 liters during the first five hours of the operation.

Cooling of Photomultipliers Directly by Liquid Nitrogen. The FEU-18 is placed in liquid nitrogen in a quartz dewar with optical windows [similar to the procedure described by Vladimirov and Litvin (1959)]. Temperature equilibrium is achieved in 15 to 20 min, and the intrinsic noise falls to 2–3 pulses/sec (Fig. 5.8). Nitrogen consumption does not exceed 2 liters during the first five hours of operation.

FEU-46A and FEU-39 cannot be placed directly into liquid nitrogen. These photomultipliers are therefore inserted into a metal chamber which in turn is placed in a thermally insulated metal envelope containing liquid nitrogen (Fig. 5.9). The end of the photomultiplier is placed against the bottom of the chamber to ensure good thermal contact.

The reduction in the dark current and in the temperature as a result of cooling of the FEU-46A placed in the chamber immersed in liquid nitrogen is shown in Fig. 5.10. Stable temperature (–188°C) is achieved 40 min after the beginning of the cooling process. The noise level is then 3–4 pulses/sec. Nitrogen consumption amounts to about 3 liters during the first five hours of operation.

End-window photomultipliers are conventiently cooled by liquid nitrogen vapor. The very simple design will ensure that practically any temperature can be achieved in a broad range between room temperature and –150°C. However, in the case of cooling below –120°C, the consumption of liquid nitrogen is quite considerable. Cooling of photomultipliers by placing them in liquid nitrogen results in the lowest noise level for minimum nitrogen consumption, but

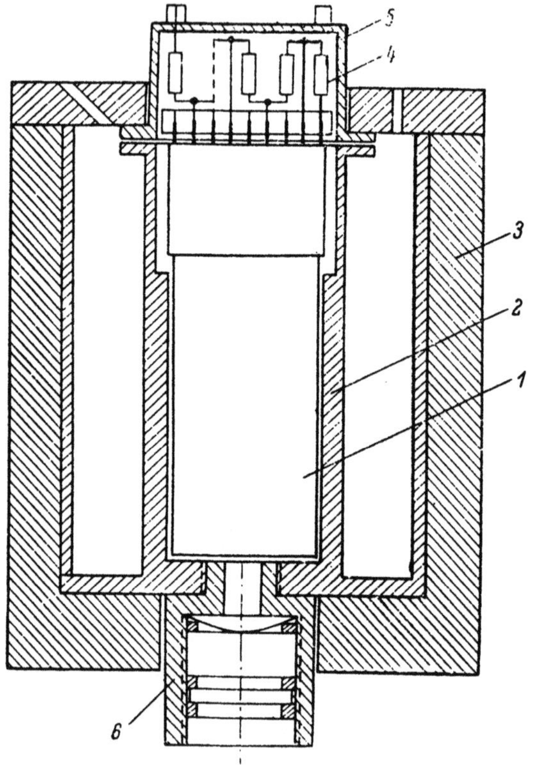

Fig. 5.9. Cooling of an end-window photomultiplier in a container. (1) Photomultiplier; (2) metal container; (3) plastic foam insulator; (4) voltage divider; (5) screen; (6) ebonite tube.

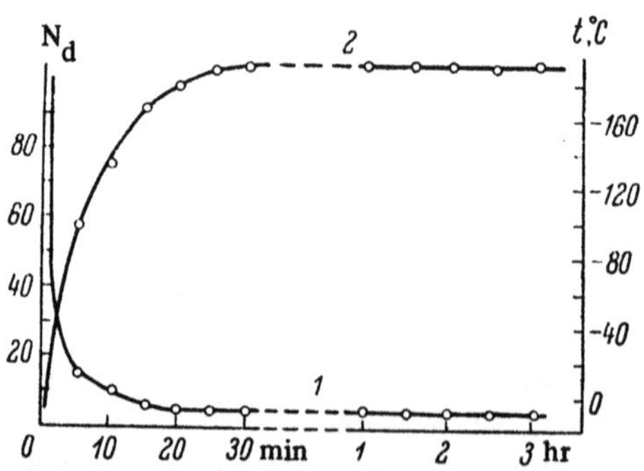

Fig. 5.10. Time characteristics of the FEU-46A photomultiplier cooled in the apparatus shown in Fig. 5.9. (1) Intrinsic noise; (2) photocathode temperature. Remaining notation as in Fig. 5.8.

this method is possible only for some photomultiplier types, and involves partial loss of light in the liquid nitrogen itself. Cooling through photocathode and dynode pins is free from this disadvantage. However, this system is more difficult to design and is suitable only for a relatively small number of photomultipliers.

Parameters of Cooled Photomultipliers. The sensitivity of the photomultiplier is of course a function of temperature. The sensitivity depends on the type of photosensitive layer, the wavelength of the incident, and, most important, on the photocathode base. Photomultipliers can thus be divided into three main groups:

1. Those in which the photocathode is placed on a massive metal backing (FEU-17, FEU-18, etc.);
2. Those in which the photocathode is deposited on glass with a semitransparent conducting base (venetian-blind photomultipliers with SbCs photocathodes, some FEU-VEI); and
3. Those in which the photocathode is deposited on glass without any backing (FEU-19, FEU-39, FEU-46, etc.).

In the case of photomultipliers belonging to the first and second groups it was found that the sensitivity does not change greatly when the photomultipliers are cooled down [Rodionov and Osherovich (1950); Osherovich *et al.* (1962); Vil'dgrube *et al.* (1963); Skachkov (1964)]. Thus, in our measurements on the FEU-57 (second group), the sensitivity change due to cooling was similar for both low and high thresholds: the sensitivity increased by a factor of 1.1 to 1.5 (Fig. 5.11) and a stable sensitivity value was established almost instantaneously. It is well known that in the case of photomultipliers belonging to the third group the sensitivity decreases beginning with –60°C, reaching a level which is lower by a factor of 50–80 at –196°C [Matveev *et al.* (1962); Osherovich *et al.* (1962)].

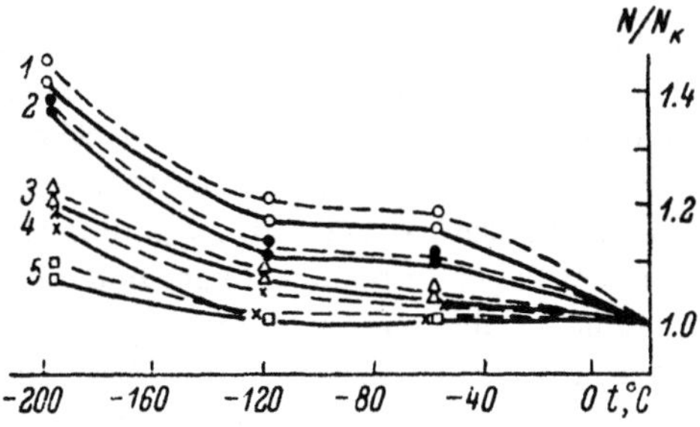

Fig. 5.11. Sensitivity of FEU-57 photomultiplier for low and high light intensities during cooling. The ratio of signals for low and room temperatures is plotted as a function of temperature. (1) 436 mμ; (2) 365 mμ; (3) 313 mμ; (4) 556 mμ; (5) 296 mμ. Solid lines – high intensity; broken lines – low intensity.

The reduction in the sensitivity of photomultipliers belonging to the third group on cooling can be explained in terms of the properties of photocells with photocathodes of the same type [Morgulis and Dyatlovitskaya (1940)]. For any point A on the photocathode we have

$$V_A = V - IR_{\text{eff}} \tag{5.1}$$

where V_A is the true potential difference between the point A and the dynode, V is the potential difference between the photocathode leads and the first dynode, I is the photocurrent, and R_{eff} is the effective resistance between A and the photocathode leads.

If V_S is the saturation potential for the collected photocurrent, then for $V_A \gg V_S$ all the photoelectrons are collected by the anode. As the temperature is reduced, the resistance R_{eff} increases, and if the light level is kept constant, the potential V_A decreases. When V_A becomes less than V_S some photoelectrons are lost and the sensitivity is reduced. Moreover, the reduction in V_A leads to a reduction in the energy of the photoelectron and consequently in a reduction in secondary emission.

It follows from Eq. (5.1) that as the photocurrent I is reduced the potential V_A will again reach V_S and the sensitivity will be restored. The current I can be reduced either by reducing the intensity of the incident radiation, or by exposing the photomultiplier to radiation of the wavelength at which the photocathode has a low sensitivity. It can be seen that for low I, the sensitivity of photomultipliers in the third group will retain its value down to very low temperatures. This is of undoubted practical interest and was confirmed by Barenboim *et al.* (1966, 1967*a*) who investigated photomultipliers belonging to the third group (FEU-39 and FEU-46A). For low light levels (5000 photons/sec) it was found that the sensitivity exhibited the same behavior at all wavelengths (apart from $\lambda = 546$ mμ in the case of FEU-46A) as the temperature was reduced. The sensitivity was in fact found to increase by a factor of 1.5 to 1.8 when the photomultiplier was cooled down to –120°C (Figs 5.12 and 5.13).

When the intensity of the incident radiation is sufficiently high (for example, 5×10^5 photons/sec, which corresponds to a photocurrent of about 5×10^{-8} A), the sensitivity is found to decrease rapidly below a certain temperature. In Fig. 5.14 the intensity corresponding to curve 2*b* is higher by a factor of 5 than for curve 2*a*. In the case of the FEU-39 photomultiplier this decrease begins at –80°C for intensities corresponding to an anode current of 5×10^{-8} A.

The constant value of the sensitivity for a given high intensity is not established instantaneously. For example, at –140°C the photomultiplier sensitivity immediately after exposure of the photocathode to the radiation corresponds to the low-intensity sensitivity at the given temperature (Fig. 5.14, curve 5*b*). It then begins to fall, reaching to its constant value (curve 5*a*) after about 2–3 min. If the exposure is terminated, the sensitivity is restored at a slower rate (30 to 40 min).

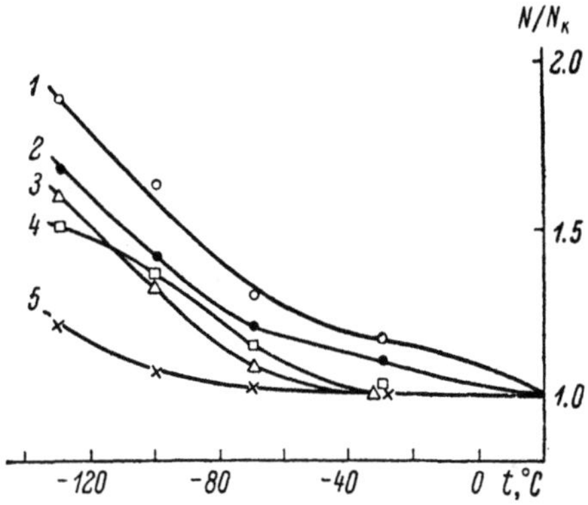

Fig. 5.12. Sensitivity of the FEU-39 photomultiplier to low intensities as a function of temperature. (1) 312.6 mμ; (2) 365 mμ; (3) 254 mμ; (4) 436 mμ; (5) 546 mμ. Remaining notation is as in Fig. 5.11.

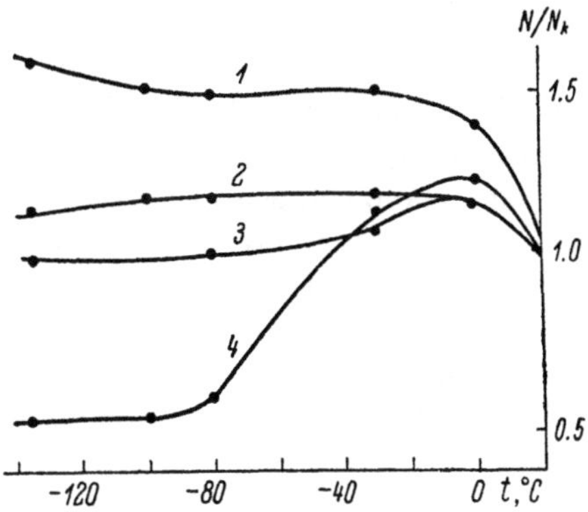

Fig. 5.13. Sensitivity of the FEU-46A photomultiplier to low-intensity radiation as a function of temperature. (1) 312.6 mμ; (2) 365 mμ; (3) 436 mμ; (4) 546 mμ. Remaining notation as in Fig. 5.11.

These slow sensitivity variations can be explained by considering the rate at which a steady state is established in the potentials of the various points on the photocathode. This rate is characterized by the relaxation time $T = \epsilon/4\pi\sigma$, where ϵ is the permittivity and σ the electrical conductivity. The photocathode resistance at room temperatures is high (the specific surface resistances is 10^7 to 10^9 ohms) and increases considerably as the temperature is reduced. There is therefore an increase in T and the steady value of the photocathode potential is established over a long period of time if a large number of electrons have been ejected from the photocathode: These phenomena have a much smaller effect on

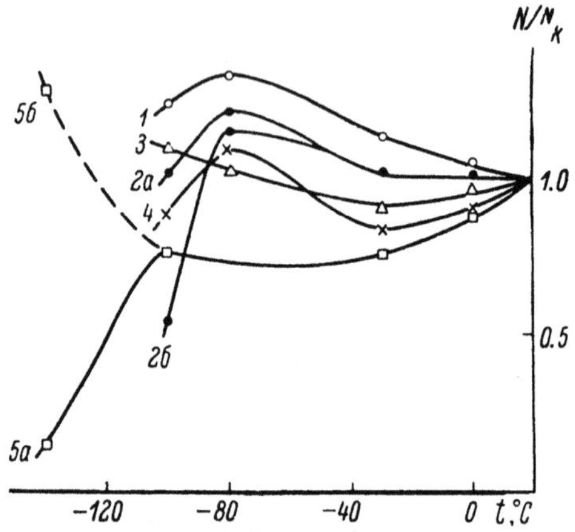

Fig. 5.14. Sensitivity of the FEU-39 photomultiplier to high intensities as a function of temperature. (1) 312.6 mμ; (2a), (2b) 365 mμ; (3) 254 mμ; (4) 436 mμ; (5a), (5b) 546 mμ. Remaining notation as in Fig. 5.11.

the low-intensity sensitivity because the potentials at the various points on the photocathode change only slightly.

In the case of cooled photomultipliers belonging to the third group, the low-intensity sensitivity remains practically unaltered, while the high-intensity sensitivity is reduced, reaching a constant level with a delay of 2–3 minutes. This means that high and low intensities cannot be measured with the same apparatus. This limitation may arise when it is desired to measure intensities which vary rapidly with time, for example, recombinational radiation, phosphorescence, or thermal luminescence of objects for which the luminescence decay time is faster than the relaxation time of the semiconducting photocathode layer. Simultaneous measurements of high and low intensities can be performed only in special cases (radiation of short duration, long intervals between measurements, and so on).

In general, photomultipliers belonging to the third group should not be cooled beyond –70 to –80°C. At such temperatures the sensitivity is practically constant, and a reduction in the dark current by two orders of magnitude yields the required increase in the signal to noise ratio.

5.3. COMPARISON OF METHODS OF RECORDING PHOTOMULTIPLIER SIGNALS

It is well known that photons incident on the photocathode eject electrons which are then multiplied at the dynodes, producing a series of pulses in the anode circuit. The signal from a photomultiplier can therefore be recorded in two different ways, namely, (1) by measuring the mean current, or (2) by counting the number of pulses.

Mean Current Measurements

Two cases are possible when the mean current in the anode circuit is measured: the output signal may be constant, or it may be amplitude-modulated (if the incident intensity is periodically interrupted, or the gain of the photomultiplier is varied). In the former case, the signal is recorded with a DC amplifier, and in the latter case it is measured with the aid of a tuned amplifier. The minimum intensity which can be recorded by these two methods is different, and it is therefore important to choose the optimum method.

Sensitivity Threshold in the Case of a Constant Output Current. The minimum light flux F_{min} which can be recorded with a photomultiplier operating in conjunction with a DC amplifier is governed by the dark current I_d and when the constant component of I_d is compensated the minimum flux is governed by the statistical fluctuations. The latter effect is connected with the fact that thermionic emission by the photocathode is essentially a statistical process, and therefore the number of electrons bombarding the anode will fluctuate in time. It can be shown that when the statistical effect is taken into account [Chechik *et al*. (1957)]

$$F_{min} = \sqrt{\frac{e(1+B)}{R_y C}} \cdot \sqrt{\frac{I_d}{\gamma_a \gamma_p}}, \tag{5.2}$$

where e is the electron charge, $1 + B$ represents the effect of the dynode system on the statistical fluctuations (in the case of photomultipliers with electrostatic focusing this quantity lies between 1.5 and 3), R_y is the insulation resistance between the anode leads and the other electrodes, C is the capacitance between the anode and the last dynode, and γ_a and γ_p are the sensitivities of the anode and the photocathode, respectively.

It is evident from Eq. (5.2) that the sensitivity threshold may be reduced by improving the insulation of the anode, by increasing the sensitivity of the photocathode, by reducing the bandwidth of recorded frequencies, and by reducing the dark current. The anode insulation and the dark current change in the required way when the photomultiplier is cooled. However, it is important to take into account not only the noise due to the dark current, but also the photocurrent fluctuations. We thus have

$$\frac{\rho_2}{\rho_1} = \sqrt{\frac{I_p + I_{d_1}}{I_p + I_{d_2}}}, \tag{5.3}$$

where ρ_1 and ρ_2 are the signal to noise ratios at room and low temperatures, I_p is the photocurrent, and I_{d_1} and I_{d_2} are the dark currents at room and low temperatures, respectively.

When $I_p \geqslant I_{d_1} \geqslant I_{d_2}$ the gain resulting from the cooling of the photomultipliers is practically zero because ρ_2/ρ_1 is then practically equal to unity. The situation is different when $I_{d_1} \gg I_p \gg I_{d_2}$. In this case $\rho_2/\rho_1 = \sqrt{I_{d_1}/I_p}$ and the gain achieved on cooling is governed by the difference between I_p and I_{d_2}. It follows that light collection is important [increase of I_p for obtaining the maximum value of ρ_2 and for the maximum cooling effect (ρ_2/ρ_1)].

The minimum light flux which can be recorded with the photomultiplier operating in conjunction with a DC amplifier corresponds to a photomultiplier output current of the order of 10^{-11} A. The DC amplifiers which are capable of recording such signals are relatively simple and do not have a large time constant or zero drift. Such amplifiers are available commercially (for example, the EMU-4, the 'Kaktus' amplifier, and so on). It is then sufficient to cool the photomultiplier down to temperatures of the order of −80°C, in which case the dark current is reduced by more than two orders of magnitude.

Sensitivity Threshold in the Case of Modulated Signals. If the light flux (or the voltage on one of the dynodes) is modulated [Sushchinskii (1950); Borisov (1960); Gavrilenko *et al.* (1962); Groves (1965)] the signal must be amplified by a tuned amplifier, followed by a synchronous detector and a meter or pen-recorder. This system has a number of advantages as compared with a DC method: (1) considerable amplification of the system can readily be achieved, (2) there is no zero drift, (3) photoluminescence, Raman scattering, and so on can be investigated by using line-operated gas-filled light sources, so that the light output is line-frequency modulated, and (4) the photomultiplier sensitivity threshold can be reduced roughly by an order of magnitude down to about 50 photoelectrons per second [Borisov (1960)]. This threshold is achieved because the effect of the constant component of the dark current is eliminated in the case of modulation. Moreover, narrow-band amplification and synchronous detection give rise to a reduction in the recorded thermionic current. This occurs because noise is a statistical phenomenon and is produced in a broad band of frequency, while the amplifier is tuned to the modulation frequency. Further reduction in the sensitivity threshold encounters fundamental difficulties. If the light flux is such that the rate at which the photoelectrons reach the detection circuit is less than the modulation frequency, the photomultiplier output signal will not always vary at the modulation frequency because the photons will not reach the photocathode in each period.

The relative root mean square fluctuation in such cases is given by [Van der Ziel (1958)]

$$\delta = \frac{1}{\sqrt{N}} \sqrt{\frac{\sigma}{\sigma - 1}}, \tag{5.4}$$

where N is the number of electrons leaving the photocathode in each modulation period and σ is the secondary emission coefficient. The condition for satisfactory recording is then $\delta < 1$. If we suppose that $\delta = 1/3$ and $\sigma = 5$, these conditions are satisfied if about 10 photoelectrons are emitted from the cathode in each half-period. For a modulation frequency of 30 Hz the minimum light flux under these conditions is $30 \times 10 \times 2 = 600$ electrons per second. It is then important to note that the modulation period is assumed to be of the same order as the time constant of the recording system.

Approximate calculations show that an increase in the modulation frequency at constant light flux and time constant of the recording device may lead to an increase in the fluctuations. This has been observed by Ksanfomaliti (1962) who recommends an optimum frequency which is determined for each photomultiplier from the plot of the fluctuation level as a function of the supply voltage and modulation frequency for a given intensity.

It may therefore be concluded that low modulation frequencies are desirable. However, a reduction in the modulation frequency leads to complications in the case of dynamic experiments. Moreover, very low modulation frequencies are inconvenient from a practical point of view. It follows that a threshold of about 50 photoelectrons per second, which corresponds to a minimum modulation frequency of a few tenths of cycles per second, is the limit for apparatus in which modulated signals are recorded. It is then desirable to cool the photomultiplier down to temperatures which reduce the dark current down to about 30 thermoelectrons per second.

If the modulation frequency is a few hundred Hz, cooling of the photomultipliers is inconvenient because the light flux for satisfactory detection is then of the order of several thousand photoelectrons per second and the intrinsic photomultiplier noise at room temperature approaches 1000 pulses per second. This is confirmed by the data reported by Ksanfomaliti (1962) who did not succeed in achieving any real advantage by detecting a light flux of this modulation frequency with a cooled photomultiplier.

A disadvantage of the above method of detection is the necessity of maintaining the modulation frequency at an accurately constant value (if a tuned amplifier is used), and the absence of mass-produced devices capable of recording a modulated photomultiplier signal.

Pulse Counting

Each electron emitted by the photomultiplier photocathode is converted at the anode into a pulse approximately 10^{-9} sec long. This consists of approximately M electrons, where M is the photomultiplier gain. In view of their origin, such pulses are commonly referred to as single-electron pulses. Since the intrinsic pulse length is very short, the pulse parameters are really governed by the time constants of the input circuits. For example, if the input capacitance is 10 pF and the load resistance is $R_L = 10\ \text{k}\Omega$, the length of the amplifier signal will be $\tau = 10^{-11} \times 10^4 = 10^{-7}$ sec, and the amplitude will be $U = (eM/\tau)R_L$ where e is the electron charge. If we suppose that $M = 10^6$, it turns out that for the above parameters, $U = 16$ mV. Amplification and counting of such pulses presents no particular difficulty and can be performed with commercially available devices. Pulse amplifiers of type USh-2 "Siren" and USh-10 "Evkalipt" are often used. Scaling units commonly employed are PS-10,000 "Floks" and PP-8 "Volna." Type ISS-3 and "Tyul'pan" rate meters, or the counting stack of type SSB, are also widely used.

Pulse counting exploits to the full the advantages of a low-noise photomultiplier. Thus, the number of pulses N which can be recorded in a time τ is determined only by the number N_b of background pulses. The relative error in the counting rate is given by [Chechik *et al.* (1967)]

$$\eta = \frac{\sqrt{N+N_b}}{N-N_b}, \tag{5.5}$$

It does not follow, however, that as N_b tends to zero the counting rate N will also tend to zero. For a very low N_b the sensitivity threshold is limited by fluctuations in the light flux. This problem has been discussed in detail by Ratner (1960). If $N_b \to 0$ for a chosen τ the sensitivity threshold will tend to a constant value. Thus, for example, the flux which can be detected in one measurement with a probability of 0.84 for a given N_b is $N = 1+\sqrt{0.75+2N_b}$. As $N_b \to 0$ we have $N = 1.85$. It follows that, for a given time of measurement, the background can be reduced only to a certain value. Further reduction in the background will have a useful result only if the time of recording is increased at the same time. This consideration is important with respect to observations of the chemiluminescence of biological objects because its intensity is very low.

The pulse-counting method gives the best results but only for some of the photomultipliers, e.g., those for which the amplitude distribution of the single-electron pulses follows the Poisson law (see Section 5.5).

The maximum light flux which can be recorded in this way is in principle limited by the speed of the scaling units, most of which cannot operate at rates higher than 10^6 pulses per second. However, in practice, the counting rate is a nonlinear function of the light flux for counting rates of the order of 10^4 pulses per second. This phenomenon can be explained both by counting losses in the scaling unit due to the statistical distribution of the incoming pulses, and also by the self-discrimination effect which will be discussed in Section 5.4.

It follows that the current-amplification and pulse-counting methods should be combined in a given apparatus if it is desired to determine light fluxes in a broad range of values. The second method requires the use of a low-noise (cooled) photomultiplier. The boundary between the two methods lies at about 10^4 pulses per sec. which corresponds to a photomultiplier anode current of 10^{-9} to 10^{-10} A. If the light flux does not exceed 50–100 photoelectrons per second the modulation method is the best and does not impose any stringent requirements on the photomultiplier parameters.

5.4. PROPERTIES OF THE PULSE-COUNTING METHOD

Individual photons reaching the photocathode give rise to single-electron pulses in the photomultiplier collector circuit. The pulse-counting method has a number of properties which are connected above all with the conditions under which the photomultipler is operating, the parameters of the pulse amplifiers, and the parameters of the scaling units.

Choice of Input Parameters. The signal amplitude U_{sig} across the photomultiplier load (which takes the form of the input capacitance C_{in} and the input resistance R_{in}) is given by

$$U_{sig} = i_{PM} R_{in} \left(1 - e^{-\tau_{PM}/R_{in} \cdot C_{in}}\right) \tag{5.6}$$

where τ_{PM} is the length of the photomultiplier current pulse, and i_{PM} is the amplitude of the pulse. The amplitude given by (5.6) reaches a maximum for given C_{in} when $\tau_{PM}/R_{in} \cdot C_{in} \to 0$ in which case $U_{sig} \to U_{sig} = i_{PM}\tau_{PM}/C_{in}$. Consequently, to ensure that U_{sig} is as high as possible, it is necessary to ensure that C_{in} should be a minimum and $R_{in} \cdot C_{in} > \tau_{PM}$.

In practice, this result is satisfied even for $R_{in} \cdot C_{in} \simeq 10\tau_{PM}$. Usually, $\tau_{PM} = 10^{-8}$ to 10^{-9} sec and $C_{in} \simeq 10$ pF. Hence

$$R_{in} = \frac{10\tau_{PM}}{C_{in}} = \frac{10 \cdot 10^{-8}}{10^{-11}} = 10^4 \ \Omega$$

Therefore, the load resistance must not be less than 10 kilohms. The upper limit for the load resistance is governed by the following factors.

1. Counting losses in the recording system when the length of the individual pulses τ is comparable with the mean interval between pulses.

2. Fluctuations due to thermal noise in the output resistor (U_m) which increase with increasing R_{in}. Since an increase in R_{in} above 10 kΩ (for constant C_{in} and τ_{PM}) does not lead to an appreciable increase in the signal amplitude and $U_m^2 = 4kTR_{in}\Delta f$, the appearance of noise with increasing R_{in} should lead to a reduction in the signal-to-noise ratio. (T is temperature.)

3. Increase in various noise levels (induced noise etc.) with increasing load resistance.

4. Counting losses due to the appearance of a voltage across the capacitance of the control grid of the discriminator tube. This rise in voltage leads to an additional shift of the discrimination level which is equal in magnitude to the constant component of the amplified pulses. Pulses with amplitudes of the order of this constant voltage are therefore missed, i.e., we have the phenomenon of self-discrimination. The number of counts lost depends on the magnitude of the constant component of the transmitted pulses, which in turn is determined by the value of the mean ratio $Q = T/\tau$, where T is the reciprocal counting rate. As Q decreases the discriminator-level shift will increase and there will be a corre-

sponding increase in the fraction of counts lost. It is clear that there is a threshold value Q_{th} above which the counting errors become negligible. Since $Q = T/\tau$ it follows that for given T it is possible to find the maximum pulse length τ_{th} (corresponding to Q_{th}) above which counting losses become appreciable. Since the pulse length is directly related to the time constant of the input circuit (usually $\tau = 2R_{in} \cdot C_{in}$), it follows that for a given counting rate $N = 1/T$ it is possible to determine the maximum value of $R_{in} \cdot C_{in}$ corresponding to Q_{th}.

Let us assume that $Q_{th} = 50$. This corresponds to a situation in which pulses with amplitudes less than one fiftieth of the mean pulse amplitude (in a single-electron amplitude distribution from the photomultiplier) will not be counted. If we suppose that $N = 10^4$ pulses per second, we have $\tau_{th} = T/Q_{th} = 1/(Q_{th}N) = 10^{-4}/50 = 2 \times 10^{-6}$ sec, and since usually $C_{in} = 10$ pF, we have

$$R_{in}^{max} \approx \frac{\tau_{th}}{2C_{in}} = \frac{2 \cdot 10^{-6}}{2 \cdot 10^{-11}} = 10^5 \ \Omega.$$

It follows that the self-discrimination effect imposes definite limitations on the maximum value of R_{in}. This resistance must be such that for given C_{in} the time constant of the input circuit $R_{in}C_{in}$ does not exceed $\frac{1}{2}\tau_{th}$ and the quantity τ_{th} is calculated from N and the threshold value of Q. Consequently, if a photomultiplier is operated under the pulse-counting conditions, the optimum parameters of the input circuit must satisfy the inequality

$$\frac{\tau_{th}}{2C_{in}} > R \geqslant \frac{10\tau_{PM}}{C_{in}} \quad . \tag{5.7}$$

Amplitude Distribution at the Photomultiplier Output, and the Use of Discrimination for Increasing the Signal-to-Noise Ratio. If we expose the photomultiplier to a constant light flux and record pulses at the output, the amplitudes of these pulses will not be equal, but will be distributed in accordance with some statistical law. If this law is different for the photopulses and the noise pulse, it is possible to use amplitude discrimination to separate them. On the other hand, if the two distributions are the same, discrimination will be of little use. Consequently, whether or not discrimination will be useful can only be decided by comparing the two amplitude distributions.

We have already noted in Section 5.3 that the intrinsic photomultiplier noise is partly due to thermionic emission and partly to field emission. It has been shown by Khlebnikov *et al.* (1962) that thermionic emission from the photocathode is a single-electron process (current of single electrons), and if the photomultiplier dark current is exclusively due to thermoelectrons from the photocathode, then its strength I recorded by a galvanometer at the photomultiplier output will be numerically equal to the calculated current expressed in terms of (1) the number of pulses N at the photomultiplier output, (2) the electron charge e, and (3) the photomultiplier gain M, i.e., $I = eMN$. In practice this relationship is satisfied for a large number of photomultiplier types (FEU-A1, FEU-42, FEU-43). For most commercially available photomultipliers, $I \gg eMN$, i.e., the noise is largely of nonthermoelectric origin.

If we determine the dependence of the number of pulses at the photomultiplier output on the supply voltage (Fig. 5.15) we can use this dependence as an indication of the quality of the photomultiplier [Khlebnikov *et al.* (1962)]. The slope of portion (1) of the curve corresponding to low supply voltages (Fig. 5.15) is due to insufficient photomultiplier gain, i.e., loss of low-amplitude pulses. The next portion of the curve (2) is the saturation region, which corresponds to the recording of each thermoelectron multiplied by the dynode system. The small slope in this section may be due to thermionic emission by the dynodes, and the development of secondary processes in the photomultiplier (gas discharge, field emission), which increase rapidly over the final portion (3) of the curve. If follows that the existence and the slope of the plateau on the counting-rate curve will also indicate the quality of the photomultiplier. Such data can be used to subdivide photomultipliers into good and poor specimens.

Thermionic emission predominates in good photomultipliers, and because of its single-electron nature, the output signal amplitudes are distributed in accordance with Poisson's law [Yanoshi (1955); Fleishman (1962); Khlebnikov *et al.* (1962); Pertsev *et al.* (1963, 1964, 1965); Artem'ev (1964); Bay and Papp (1964); Eberhardt (1964); Hayman *et al.* (1964); Baldwin and Friedman (1965)]:

$$p_N = \frac{k^N}{N!} e^{-k}, \tag{5.8}$$

where k is a number approximately equal to the secondary emission coefficient σ_1 of the first dynode (in good photomultipliers $\sigma_1 \geqslant 5$).

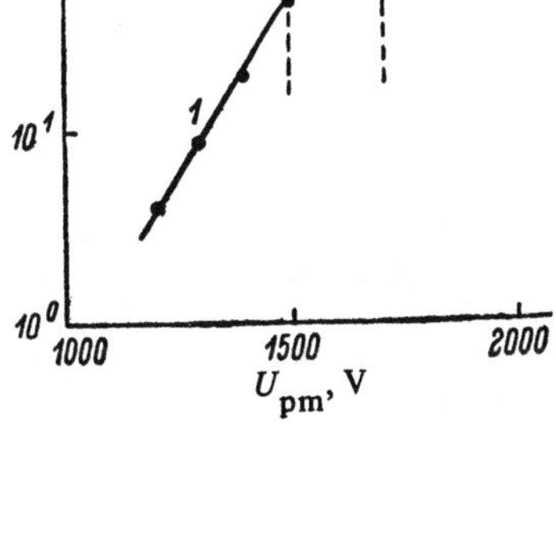

Fig. 5.15. Counting characteristic of a photomultiplier. Number of pulses per sec is plotted as a function of applied voltage.

For smaller σ_1 the Poisson distribution affects the subsequent dynodes. This leads to a considerable distortion of the initial distribution, and Fleishman (1962) has shown that in this case

$$k = \frac{\sigma_2(\sigma_1 - 1)}{\sigma_1}, \tag{5.9}$$

where σ_2 is the mean secondary emission coefficient of the dynode surfaces beginning with the second dynode.

Since the coefficients σ_1 and σ_2 depend on the interstage voltage, it follows that in photomultipliers for which the recommended voltage between the photocathode and the first dynode is greater by a factor of 1.5–2 than the subsequent interstage voltages, σ_1 and σ_2 will also be different.

In most of the papers which we have cited, a study was also made of the amplitude distribution of photoelectron pulses (for very low illumination of the photocathode), and it was shown that it has the same form as the amplitude spectrum of the thermoelectrons (the position of the peak was unaffected). An exponential output-pulse amplitude distribution is observed for "poor" photomultipliers. This may be a consequence of the fact that (1) pulses which are not of thermionic origin predominate at the photomultiplier output, (2) the photomultiplier was investigated under unsuitable conditions, or (3) the value of k was low owing to the fact that the first dynode was missed by the photoelectrons. The last effect is particularly important for venetian-blind photomultipliers [Khlebnikov *et al.* (1962); Pertsev *et al.* (1964)]. It is well known that for k approaching unity, the Poisson and exponential distributions are very similar.

Good photomultipliers are quite rarely encountered in practice. Most photomultipliers do not have a pure Poisson amplitude distribution, but a distribution of the form shown in Fig. 5.16 where a relatively large number of pulses is found in the region of low amplitudes. This can be related to thermionic emission by the first few dynodes. For such photomultipliers, good results are obtained by recording the pulses at the maximum of the single-electron peak, using differential or integral threshold discriminators. This method has been used by Kaznacheev *et al.* (1965*b*) and Kovaleva *et al.* (1965) who were concerned with very low intensities.

When discrimination is used, the best result is obtained with cooled photomultipliers. Cooling leads to a reduction in the number of thermoelectrons, i.e.,

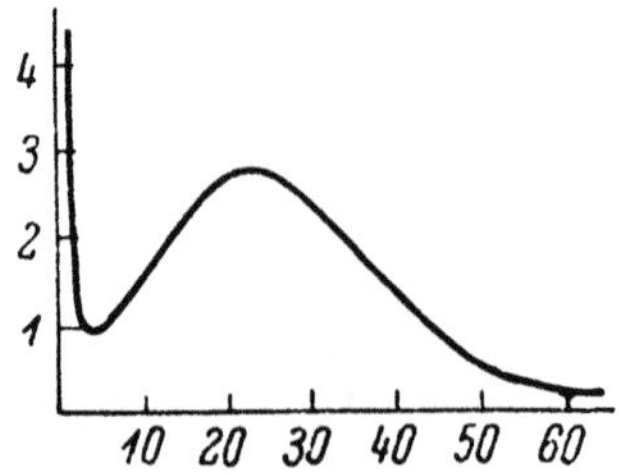

Fig. 5.16. Amplitude distribution of photomultiplier noise [FEU-15; Pertsev *et al.* (1965)]. The number of pulses is plotted as a function of pulse height in relative units.

a reduction in height of the single-electron peak [Pertsev *et al.* (1963, 1964)], which yields a significant gain in the signal-to-noise ratio for a suitably chosen temperature (see Section 5.3).

As regards the recording of very low intensities with photomultipliers having an exponential pulse-height distribution, the usefulness of discrimination is best established by comparing signal and noise amplitude distribution. Our laboratory [Novikov and Yangibaev (1966)] determined the amplitude distributions as follows. The number N_d of dark pulses is first determined for different photomultiplier voltages as a function of the amplifier gain K. These measurements yield the amplification factor K_0 above which the number of recorded pulses is independent of K. The function $N_d = f(K)$ should have a saturation region. The values of N_d corresponding to this region are used to establish the counting characteristic of the photomultiplier. The working voltage U_w is chosen to lie on the plateau of the counting characteristic. If there is no plateau (or point of inflection) it is best to use the nominal voltage for the photomultiplier. In both cases the working amplification coefficient K_w must be such that $K_w \geqslant K_0$. The pulse height distribution is then obtained for the photomultiplier at the chosen U_w and K_w.

As an example, consider the results obtained for FEU-A1 (Fig. 5.15). It is clear that for a working voltage of 1500–1600 V all the thermoelectrons emitted by the photocathode are collected; discharge phenomena are of a minor importance. The working voltage for this photomultiplier was therefore chosen to be 1600 V.

The amplitude distribution of noise and useful signal pulses (λ = 479 mμ) is shown in Fig. 5.17. It is clear that the amplitude distribution of the dark and photo pulses are the same, and agree with the theoretical Poisson distribution for k = 3.2. A second maximum appears when the number of incident photons is large. This can be related to the formation of multielectron pulses [Artem'ev (1964)]. The amplitude distribution of the photopulses due to UV radiation (λ = 260 mμ) is similar to the distribution seen in Fig. 5.17 and corresponds to k = 3 (Fig. 5.18).

Cooling of photomultipliers down to −78°C changes the shape of the distribution curve for the photopulses: the maximum disappears and the distribution is described by a theoretical curve with $k \simeq 1$ (Fig. 5.19). This effect is not in agreement with published data [Pertsev *et al.* (1963)], and may be connected with the change in the operating conditions, or in the photomultiplier parameters on cooling [Barenboim *et al.* (1966)].

It follows from the above data that the use of amplitude discrimination yields no real gain in the signal-to-noise ratio for the photomultiplier which was investigated because the signal and noise amplitude distributions are very similar.

In conclusion we must emphasize once again that, so far, we have considered only the properties of the main elements of the apparatus for the detec-

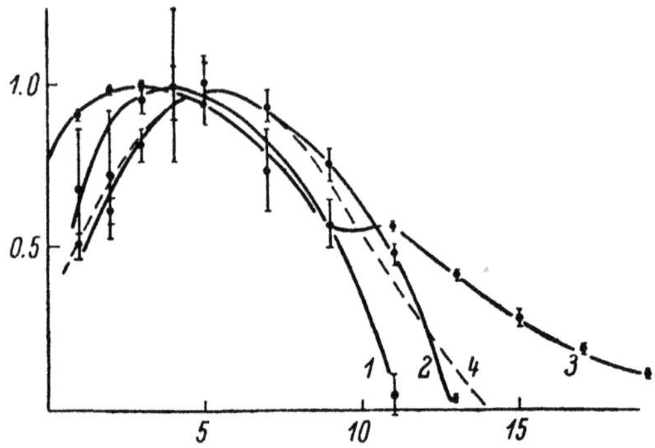

Fig. 5.17. Amplitude distribution of noise and photopulses from FEU-A1 (λ = 479 mμ). (1) Noise; (2) signal for N_0 = 224 pulses/sec; (3) signal for N_0 = 6000 pulses/sec; (4) theoretical curve for k = 3.2 (N_0 is the counting rate without discrimination). The remaining notation is as in Fig. 5.16.

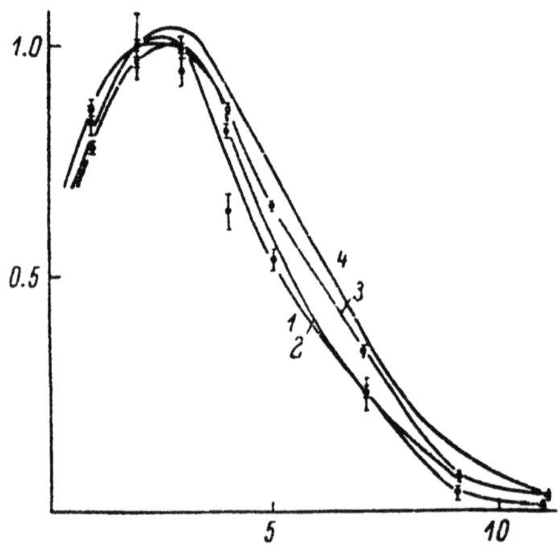

Fig. 5.18. Amplitude distribution of noise and photopulses (FEU-A1, λ = 260 mμ). (1) Noise; (2) signal for N_0 = 2400 pulses/sec; (3) signal for N_0 = 300 pulses/sec; (4) theoretical curve. Remaining notation as in Fig. 5.16.

tion of ultraviolet and weak or ultraweak radiation. The use of these properties in practice will be described in Chapter 6 in connection with special cases. It is also important to note that the concepts of weak and ultraweak radiation are interpreted in different ways by different authors. Thus, for example, in his monograph Vasil'ev (1965*c*) refers to an intensity of 20–50 photons per second as weak while Fish and others refer to intensities of 1000 photons per second as ultraweak. It would be desirable to measure the light flux in absolute units, but there are a number of practical difficulties inherent in absolute photometry. It is

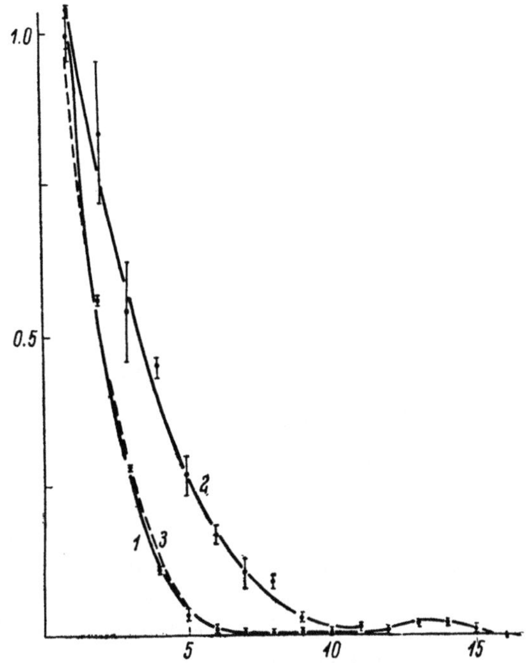

Fig. 5.19. Amplitude distribution of photopulses for a cooled FEU-A1 photomultiplier (λ = 479 mμ). (1) Signal for N_0 = 1025.5 pulses/sec; (2) signal for N_0 = 220 pulses/sec; (3) theoretical curve for k = 0.8. Remaining notation as in Fig. 5.16.

much better to classify the light flux in relation to the photomultiplier parameters used in each particular case. Ultraweak radiation can then be defined as that for which special measures have to be taken to reduce the photomultiplier noise, i.e., $I_p < I_d$ (see Section 5.3). When $10I_d \geqslant I_p \geqslant I_d$, such fluxes can be regarded as weak, and when $10I_d < I_p$ they can be regarded as strong. However, I_d depends on each particular photomultiplier. It may be assumed that the mean value is I_d = 1000 pulses per second and the light flux classified on this basis using the above criteria.

Chapter 6

APPARATUS FOR THE INVESTIGATION OF THE LUMINESCENCE OF BIOPOLYMERS AND CELLS

Progress in the investigation of the luminescence of biopolymers and cells is closely related to the development of new types of apparatus. In this chapter we shall therefore review some of the published work in this field, and also a number of relatively simple devices which we have developed for the investigation of photo-, chemi- and radioluminescence of biopolymers and cells.

In addition to construction problems there are also methodological problems, but, owing to lack of space, these problems are not considered in the present book. Some aspects of this subject have been discussed in the literature [Teale and Weber (1957); Parker and Rees (1960); Burshtein and Vladimirov (1964); Vladimirov and Litvin (1964); Parker (1964)].

6.1. APPARATUS FOR THE RECORDING OF PHOTOLUMINESCENCE

General Requirements

The photoluminescence of biological specimens imposes certain specific requirements on the apparatus employed to detect it. An optimum apparatus must have the following properties: (1) universality, (2) high sensitivity to weak and ultraweak light fluxes, (3) wide working range, and (4) stability.

When one is concerned with energy migration in proteins, it is necessary to know the excitation spectra, the luminescence spectra, and the polarization spectra [Weber (1960*a*)]. In studies of singlet–triplet transitions in proteins and amino acids in the presence of oxygen, one is concerned with the fluorescence and phosphorescence spectra, and also the phosphorescence time constant [Barenboim (1963*a*, 1963*b*)]. Universality, which means the ability of the apparatus to record a number of luminescence parameters for a given specimen, is therefore a necessary quality in such (and similar) measurements. High sensitivity, including the ability to detect individual photons, is necessary for a number of reasons, including: (1) the low quantum yield of many materials–the quantum yield of thymine, for example, does not exceed 0.015 under optimum conditions [Udenfriend and Zaltzman (1962)], (2) rapid photochemical destruction of biopolymers or suspensions of living cells impose a definite limit on the intensity of the exciting radiation, and (3) biologically important materials are often difficult to obtain in large enough quantities, which means that it is sometimes necessary to work with very small volumes and concentrations.

A broad dynamic range is also a necessary condition. While in tryptophan-containing proteins the quantum yield may reach 10–15% [Burshtein (1961)] in some proteins in which the main chromophore is the tyrosine group, the quantum yield does not exceed fractions of a percent [Weber (1961*a*)]. The quantum yield varies widely in kinetic experiments. Thus, when the pH of the medium is reduced the integrated fluorescence intensity of DNA increases by two or three orders of magnitude in comparison with the figure of 10–100 photons per second for neutral pH [Barenboim *et al.* (1963*b*)].

It should be noted that Raman emission of the solvent and measuring-cell luminescence interferes with the utilization of high-sensitivity apparatus. For aqueous solutions Raman peaks at 280, 309, 329, and 350 mμ correspond to the mercury lines and 256, 280, 296, and 313 mμ. The fluorescence spectrum of quartz cells has a sloping form with a broad maximum in the region of 400–420 mμ. The fluorescence intensity of the 10 × 10 mm quartz cells of the SF-4 spectrophotometer on excitation with a wavelength of 280 mμ is equivalent to the luminescence of an aqueous solution of tryptophan in a concentration of about 5×10^{-7} *M*. The intensity of the Raman lines is of the same order.

High stability is particularly important because the luminescent characteristics of biological objects exhibit a considerable spread because these objects are highly inhomogeneous and their state is variable. This means that a large number of experiments must be carried out in order to obtain statistically significant results.

Moreover, an apparatus for the detection of the photoluminescence of a biological specimen should preferably have a facility for the automatic recording of corrected spectra, and should operate at a sufficiently high rate. Finally, simplicity and low cost are also quite important.

Some Features of Apparatus for the Detection of Photoluminescence

We shall now discuss some of the measures which have to be taken in the design of apparatus in order to fulfill the above requirements. The excitation spectra are frequently corrected by the Parker method [Parker (1958)]. The apparatus incorporates two photomultipliers one of which intercepts the luminescence from the objects under investigation, and a proportion of the exciting radiation is reflected by a quartz plate onto the other photomultiplier. The signal A_1 from the photomultiplier recording the luminescence is determined by the intensity of the exciting monochromatic radiation S, and the dependence of the photocurrent on the supply voltage $f_1(U)$ and the intensity I of the object, i.e.,

$$A_1 = K_1 \cdot S \cdot I \cdot f_1(U), \tag{6.1}$$

where K_1 is a constant.

The control signal is determined by the intensity S of the exciting radiation, the dependence of the photocurrent on the supply voltage, $f_2(U)$, and the spectral sensitivity of the photomultiplier, $f(\lambda)$, i.e.,

$$A_2 = K_2 \cdot S \cdot f_2(U) \cdot f(\lambda). \tag{6.2}$$

If we divide one signal by the other we have

$$\frac{A_1}{A_2} = \frac{K_1}{K_2} \cdot \frac{f_1(U)}{f_2(U)} \cdot \frac{1}{f(\lambda)} \cdot I. \tag{6.3}$$

It follows from this formula that if the two photomultipliers are identical $[f_1(U) = f_2(U)]$, and the effect of (λ) is excluded (this is usually achieved by placing a screen in front of the photomultiplier), the signal ratio is linearly related to the change in the luminescence intensity from the object under investigation for a given wavelength of the exciting radiation. The two signals can be divided electromechanically with the aid of an automatic electronic potentiometer [Mekshenkov and Andreitsev (1960)]. Some workers [Eisenberg *et al.* (1962)] have used the control signal for automatic amplifier gain control, while others [Chechik *et al.* (1957); Borisov and Mokhova (1964)] have used photomultiplier gain control achieved by applying a control signal to the grid of a tube connected in series with the PM power supplies.

The above principle for obtaining corrected excitation spectra is widely used. Among the instruments using this principle are the universal recording device described by Mekshenkov and Andreitsev (1961), the double-beam spectrofluorimeter described by Melhuish and Murashige (1962), and the spectrofluorophosphorimeter reported by Haugen and Marcus (1964). Ainsworth and Winter (1964) have developed an apparatus which is capable of recording automatically the degree of polarization of fluorescence as a function of the wavelength of the exciting radiation, as well as the absorption, excitation, and fluorescence spectra corrected by the Parker method.

A still more sophisticated instrument has been described by Rosen and Edelman (1965). This is a stabilized spectrofluorimeter capable of recording corrected excitation spectra, and suitable for the investigation of fluorescence and phosphorescence in polarized light as functions of the temperature and the pH of the specimen.

The apparatus described above is not suitable for the recording of ultralow intensities, since the sensitivity threshold is limited by the photomultiplier noise at room temperature. Universality and high sensitivity are combined in the apparatus developed at the Biophysics and Isotope Laboratory of the Academy of Sciences of the Belorussian SSR [Katibnikov and Konev (1961); Bobrovich *et al.* (1965)]. The apparatus is capable of measuring the fluorescence spectra, the effect and the degree of polarization at room and low temperatures, and the spectral characteristics of phosphorescence with decay times as low as 10^{-3} sec. However, the apparatus does not incorporate attachments eliminating variations in the source intensity, and therefore the spectra obtained with it must be corrected for the spectral sensitivity of the photomultiplier and the dispersion of the monochromator, i.e., it does not have the advantages listed for the apparatus described above. Kovaleva *et al.* (1965) described a high-sensitivity instrument which can be used to record the spectra of some biopolymers. The luminescence excitation and the determination of its spectral composition is carried out with the aid of the DMR-4 monochromator. A special reversing scaling unit is used to subtract automatically noise levels from the useful signal. However, the absence of universality, stability, and automatic spectrum correction is characteristic even for this apparatus.

It follows that none of the above instruments satisfy simultaneously all the requirements listed at the beginning of this section, although in some cases this is justified by the fact that the apparatus is designed for some special measurements. It is clear from this brief review that the Parker method is used quite widely for stabilization and correction, while the pulse-counting method, using low noise photomultipliers, is utilized as a means of increasing the sensitivity. A universal apparatus combining both systems would largely satisfy the necessary requirements. This hybridization is achieved in the apparatus described below [Barenboim *et al.* (1964*a*, 1967*c*); Domanskii *et al* (1966)].

6.2 A UNIVERSAL PHOTOELECTRIC APPARATUS

The apparatus was designed for the recording of fluorescence, phosphorescence, absorption and excitation spectra, as well as their polarizations, in the visible and ultraviolet spectral regions, over a broad range of intensities (Fig. 6.1).

Block Diagram. White light from the source 2 (Fig. 6.2*a*) passes through the diaphragm 3, condenser 4, and double monochromator 6 (type DMR-2). Lenses 7 and 11 focus it on the front plane of the specimen cell 12. Some of the light (4%) is reflected by the quartz plate 8 on to a uranium glass placed in front of the photomultiplier in the reference circuit.

The luminescence is collected by lenses 16 and 19. It is transmitted through a ZMR-3 monochromator 20 and is focused on the photocathode of the FEU photomultiplier 22. Whenever necessary, optical filters can be inserted in front of the chamber 14. When the phosphorescence spectrum is determined (Fig. 6.2*b*) the exciting radiation reaches the specimen 26 through apertures in the phosphoroscope drum 24 and the optical window of the dewar 25. Phosphorescence is collected by lenses 16 and 19 on the slit of the ZMR-3 monochrom-

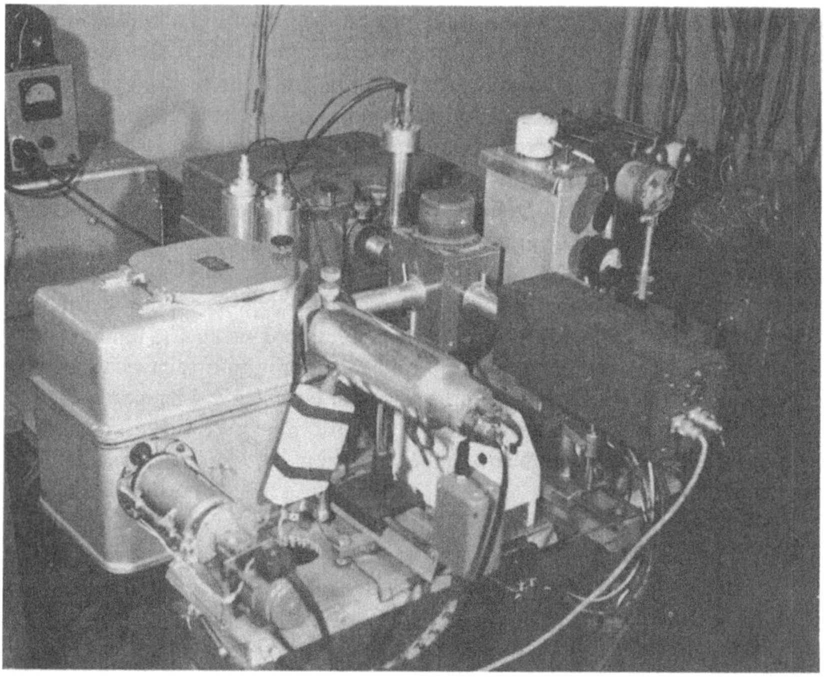

Fig. 6.1. Photograph of the optical and mechanical parts of the universal photoelectric setup.

ator after passing through the optical windows of the quartz dewar and the aperture in the phosphoroscope drum. When absorption spectra are measured the specimen cell 12 is rotated at right angles to the incident radiation. The transmitted light is recorded by a second photomultiplier. The polarization spectra and the intensity are recorded with the aid of polaroids 10, 15, and 17. These polaroids can be fitted in two fixed mutually perpendicular positions.

The light flux is recorded by photomultipliers. Signal from the photo tube 22 is first amplified (USh-2 or 'Kaktus' amplifier) and is divided by the signal from the reference photomultiplier 9. The result is recorded by the automatic potentiometer [ratio recorder] EPP-09M2.

Let us now discuss the various elements of the apparatus in greater detail.

Optical-Mechanical Part. The lens system ensures sufficiently effective light collection. The lenses are made of quartz subjected to special thermal treatment which reduces its intrinsic fluorescence. Rotation of the prisms in both monochromators is achieved with the aid of a reversing motor through reduction gears. At maximum speed of rotation of the prisms the fluorescence spectrum in the region 300–400 mμ is recorded in 32 sec. The speed can also be reduced by factors of 4, 16, and 64. Wavelength marking is produced automatically on the chart of the EPP-09 pen recorder. The prisms are brought automatically to rest after a given wavelength interval has been traversed.

The degree and the spectrum of the polarization are measured with the aid of polaroids. The polaroid holder (Fig.6.3) ensures their rotation through 90°. The polaroid film is inserted into the holder 2 and is oriented in the required fashion relative to it. The system is then inserted into the slot in the tube 5 and is covered by the lid 3. The whole assembly is made rigid by means of the sliding tube 4. Polaroids placed in the special holders can be readily removed if necessary or replaced by other polaroids.

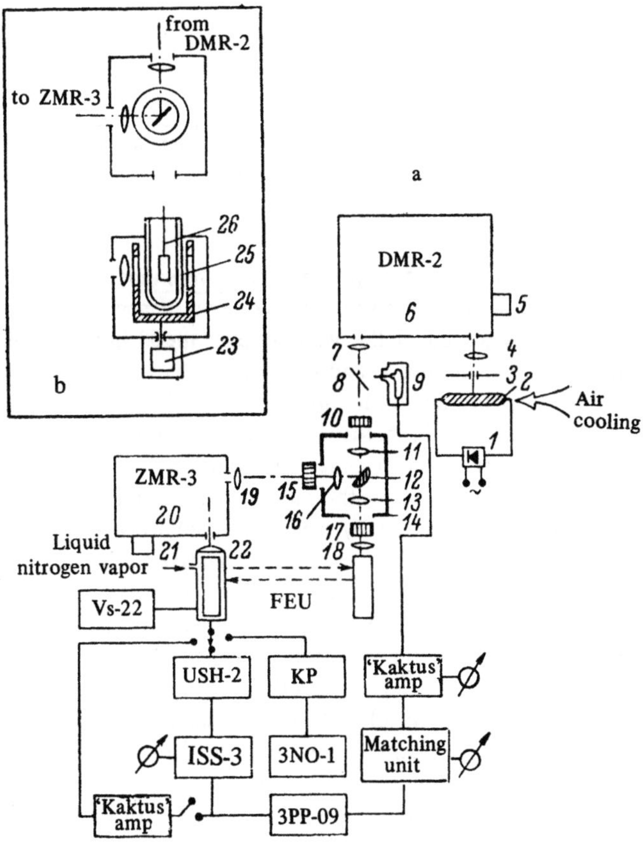

Fig. 6.2. Block diagram of the universal apparatus (a) and low-temperature attachment (b). (1) Power supply for lamp; (2) lamp; (3) diaphragm; (4) condenser; (5), (21) RD-09 motors; (6) DMR-2 monochromator; (7), (11), (13), (16), (18), (19) quartz lenses; (8) semitransparent plate; (9) reference photomultiplier; (10), (15), (17) polaroids; (12) container; (14) light-tight chamber; (2) ZMR-3 monochromator; (22) cooled photomultiplier; (23) DVA-U4 motor; (24) phosphoroscope drum; (25) quartz dewar; (26) specimen-holder.

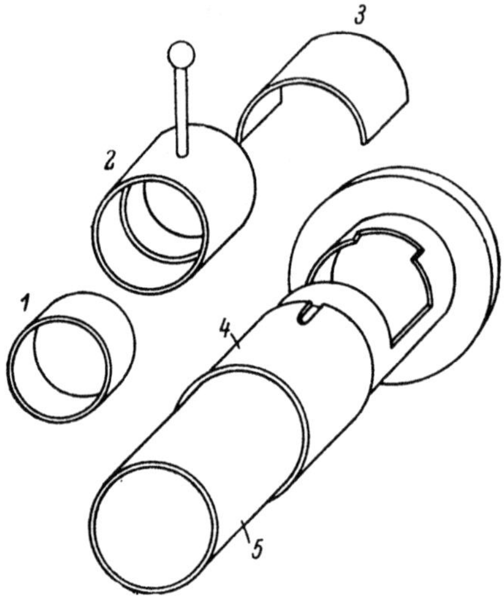

Fig. 6.3. Polaroid holders. (1) Fixing ring; (2) polaroid holder; (3) lid; (4) slider; (5) tube.

The universal thermostated specimen cells were made in two versions. The first version (Fig. 6.4*a*) is used to record the fluorescence and absorption parameters. The cell 1 is placed in the water-cooled metal holder 3 which can be rotated relative to the support 4, and can occupy two fixed positions one perpendicular to the light beam and the other at 30° to this beam.

The second version (Fig. 6.4*b*) is used when a number of specimens are to be compared in a short space of time. The solutions are poured into the four semicylindrical slots 2, covered by the quartz plates 3 and lids with windows 4. Thermal stabilization is achieved by means of circulating water fed in through the pipes 1. The whole system can be rotated relative to the support 6, and can be set in four fixed positions (by means of the screw 8).

Sources. Provision is made in the apparatus for interchangeable light sources. Absorption, excitation, and luminescence spectra are measured with high power lamps producing a continuous spectrum (DKsSh-200 and DKsSh-1000). These lamps are placed in a special metal envelope and can be adjusted relative to the monochromator slit. The SVD-120A mercury lamp is also used in the recording of luminescence spectra.

In most cases, phosphorescence is investigated with the aid of a phosphoroscope, using continuous excitation from a lamp whose output is modulated by a rotating disc or the phosphoroscope drum. In the case of biopolymers and cells it is frequently desirable to use single-shot excitation because the well-defined phosphorescence spectra of these objects change their form very rapidly as a result of photolysis. The principle of single-shot excitation has not as yet been adequately developed [Kislyak (1958); Pilipovich and Sveshnikov (1958);

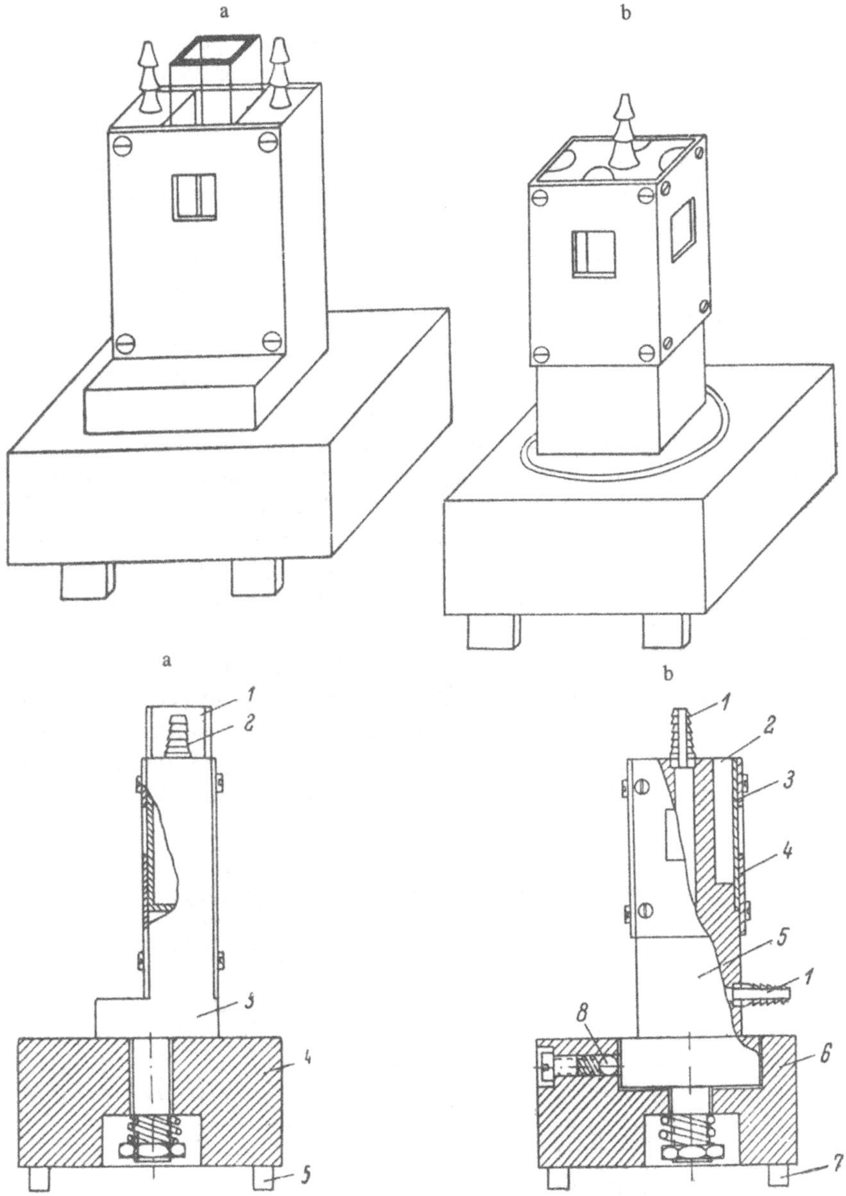

Fig. 6.4. Thermal stabilization of the sample container. (a) Fluorescence and absorption measurements: (1) quartz specimen container of a spectrophotometer; (2) connecting pipe; (3) body of container; (4) stand; (5) pins. (b) Relative studies of the fluorescent characteristics of several specimens: (1) connecting pipe; (2) slot for inserting specimen; (3) quartz window; (4) frame; (5) body of container; (6) stand; (7) pins; (8) fixing screw.

Kaminskii (1959)]. The DRSh-50 lamp, operating under pulsed conditions (see Section 5.1), is used in the above apparatus for single-shot excitation of phosphorescence.

Circuits. The circuits used in this apparatus ensure high sensitivity, operation under both pulse-counting and current-recording conditions, and division of the main signal by the reference signal thus ensuring high stability and correction of the results.

All the luminescence characteristics are measured with the aid of the FEU-39 photomultiplier cooled with liquid-nitrogen vapor down to the optimum temperature for the given incident intensity, but not less than $-150°C$ [Barenboim *et al.* (1966)]. The absorption spectra are measured with the aid of the FEU-57 photomultiplier which has peak sensitivity at 280 mμ. The photomultiplier signal is then applied to the dividing circuit.

The dividing circuit can be used in either current or pulse-counting modes. In the former case the measured signal is fed into the input of a DC amplifier and then on to the automatic potentiometer EPP-09. When the individual pulses are counted the signal is amplified by a wide-band amplifier and is integrated by the ISA-3 unit. It is then fed into a matching unit which couples the recording and standardizing channels, and then on to the EPP-09. In both cases the standardizing channel consists of the FEU-39 photômultiplier and a 'Kaktus' amplifier. The output from the amplifier is fed into the pen recorder EPP-09.

The entire recording and dividing system consists of generally available 'Kaktus,' 'Siren,' ISS-3, and EPP-09M2 units. The simple matching unit shown in Fig. 6.5 is the only addition. This unit is taken directly off the chassis of the 'Kaktus' amplifier and is supplied directly by the 'Kaktus' circuit.

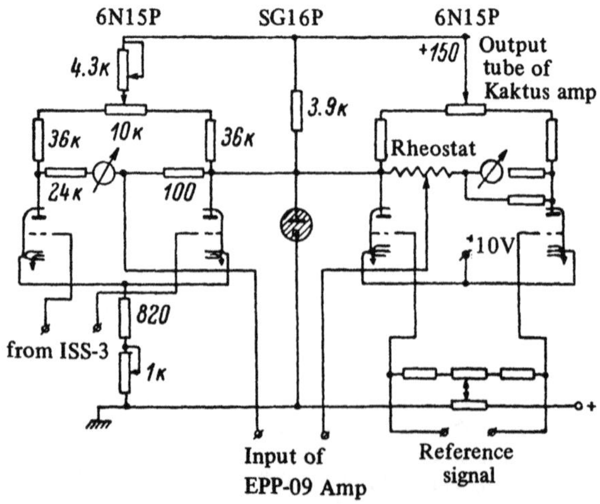

Fig. 6.5. Basic circuit of the matching unit.

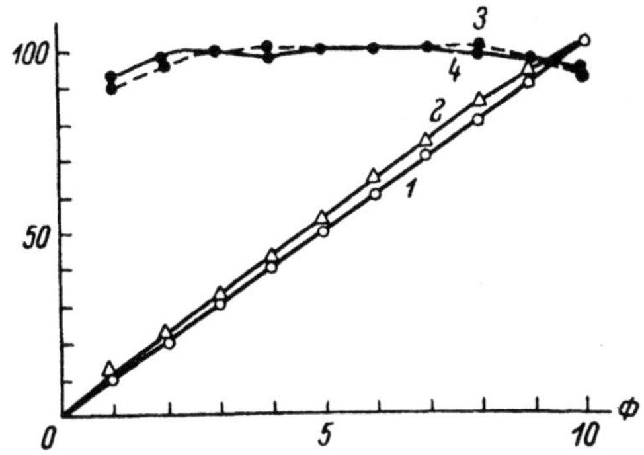

Fig. 6.6. Characteristics of the dividing system. Readings of the recording devices (relative units) as a function of the light flux (relative units). (1) Signal in the reference channel; (2) signal in the measurement channel; (3) calculated curve obtained by dividing the signal in the measurement channel (curve 2) by the signal in the reference channel (curve 1); (4) readings of the ÉPP-09M2 recorder.

Figures 6.6 and 6.7 illustrate the operation of the dividing system. It is clear that dividing errors lie in the region of small signals. This is due to instability in the standardizing circuits of the DC amplifiers when the signals from them are applied to the EPP-09M2 pen recorder.

When the phosphorescence decay constant is recorded by the single-shot excitation method, the signal from the photomultiplier is fed through a cathode follower into a type ENO-1 oscillograph with a long-persistence screen. The lifetime is determined photographically, or visually, as the time corresponding to a reduction in the phosphorescence intensity by a factor of *e*. For the accurate determination of the lifetime it is important to keep the point corresponding to the onset of the phosphorescence decay in the same position on the oscillograph screen. When this is arranged, one can use the attachment shown schematically in Fig. 6.8. A photoresistor 7 is mounted on the oscillograph screen 6 and controls the flash emitted by the lamp 1. When the cathode-ray beam reaches the photoresistor this triggers a circuit which in turn fires the lamp. An important addition to the circuit is the logarithmic amplifier which produces on the oscillograph screen a straight line when the phosphorescence decay is exponential. This is of interest in the analysis of phosphorescence mechanisms.

6.3. A SIMPLE SPECTROFLUORIMETER

In some luminescence studies it is sufficient to determine the fluorescence intensity and spectra emitted by the object under investigation (see Chaps. 1 and 2). In such cases the universal apparatus described above is inconvenient. Below we describe a simple spectrofluorimeter designed for the determination of the fluorescence intensity and spectra of solutions and cell suspensions.

A Simplified Spectrofluorimeter

This version of the instrument is interesting because it incorporates only commercially produced units which do not require additional adjustment or any serious modification.

Block Diagram. The light source is an OI-18 lamp 1 (Fig. 6.9) supplied from the line through a regulator 2. The exciting radiation is defined by a combination of light filters 3 (gaseous chlorine–bromine and UFS-1). The gas filter can be replaced by the more readily available liquid filter (see Section 5.3). The exciting radiation is focused by the lens 4 on the quartz container 5 which holds the sample under investigation. The container is inserted into an ebonite light-tight chamber attached to the monochromator 7 (ZMR-3) and the lamp. The luminescence is focused by the lens 6 on the entrance slit of the monochromator, and after resolution into a spectrum falls on the FEU-39 photomultiplier 8. The photomultiplier is supplied by a stabilized 'Orekh'-type voltage source 9 and is placed in a container incorporated in the supply source.

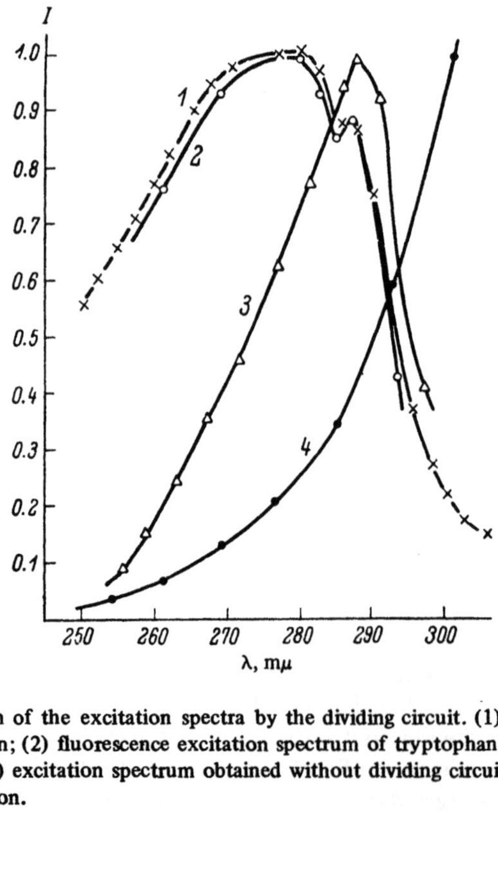

Fig. 6.7. Correction of the excitation spectra by the dividing circuit. (1) Absorption spectrum of tryptophan; (2) fluorescence excitation spectrum of tryptophan obtained with the dividing circuit; (3) excitation spectrum obtained without dividing circuit; (4) spectrum of the exciting radiation.

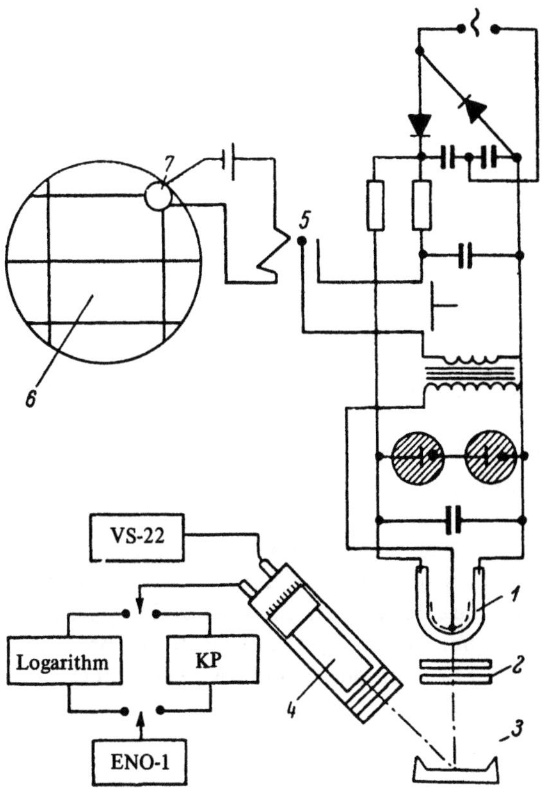

Fig. 6.8. Attachment for phosphorescence time constant measurements. (1) Pulsed lamp; (2) light filters; (3) specimen container; (4) photomultiplier; (5) RP-5 relay; (6) oscillograph screen; (7) FS-K1 photoresistor.

The photomultiplier signal is fed into a DC amplifier taken from the 'Kaktus' X-ray microtube 10, and is recorded by the EPP-09 automatic potentiometer 11. The motor 12 (SRD-2), which rotates the monochromator prism, discharges a 0.1– μF capacitance during each rotation into the input of the automatic potentiometer. This produces pulses on the pen recorder strip which are used as reference lines. Instead of the 'Kaktus'–EPP-09 combination one can also use the automatic potentiometer EPP-60, which has a high input impedance.

Improved Spectrofluorimeter

In contrast to the above simple instrument, the improved version of the device is capable of handling low-intensity samples at room and low temperatures. The block diagram of the apparatus is, in principle, the same as that shown in Fig. 6.9 but there are certain important modifications.

Luminescence Excitation. The light source is OI-18 but the output of the SVD-120 lamp is stabilized in intensity by an optical feedback system (see Section 5.1) in the part of the spectrum used for the luminescence excitation.

Detection. No special measures are taken to reduce the photomultiplier noise but the intensity is increased by selective amplification of a modulated useful signal. This is achieved by modulating the voltage on one of the dynodes of the FEU-39 photomultiplier by the method described by Borisov (1960). The alternating component of the signal is then amplified by a tuned amplifier, detected by a synchronous detector, and finally fed into the automatic potentiometer EPP-09. This system can record a usable signal as small as 10^{-11} A from the photomultiplier without any special measures for reducing the intrinsic photomultiplier noise.

Low-Temperature Measurements. The sample is placed in a hermetically sealed container with a cooling attachment and a thermocouple which measures directly the temperature of the material under investigation (Fig. 6.10). To prevent frost formation the upper window of the chamber is exposed to hot air from an annular distributor lying along the perimeter of the window. This method has been found to be more effective than exposure of the upper window to dry cold air, rubbing with alcohol, or heating. The design of the container is such that it is possible to evacuate it or to produce a positive pressure of 0.5 atm.

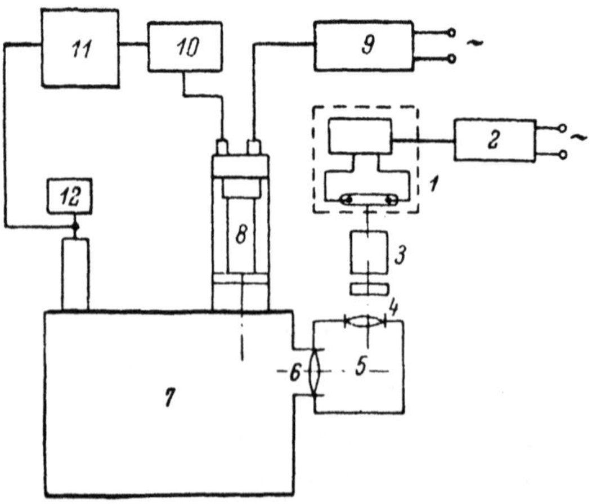

Fig. 6.9. Block diagram of a simple spectrofluorometer. (1) OI-18 lamp; (2) SM-500 stabilizer; (3) light filters; (4) (6) lenses; (5) specimens under investigation; (7) ZMR-3 monochromator; (8) photomultiplier; (9) high-voltage source; (10) DC amplifier; (11) EPP-09 recorder; (12) SRD-2 motor.

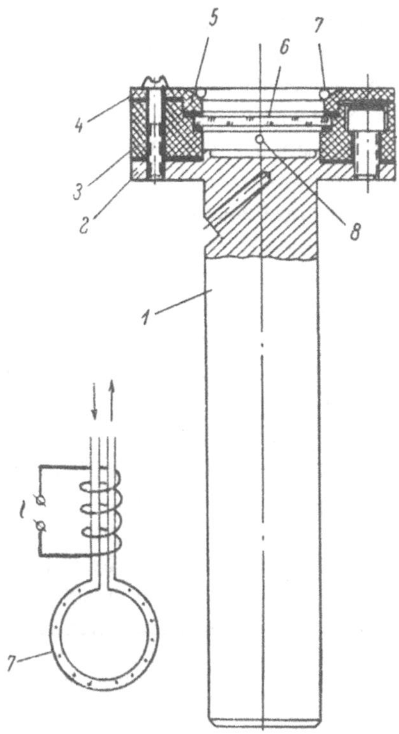

Fig. 6.10. Specimen container for low-temperature measurements. (1) Heat conductor; (2), (5) seals; (3) textolite insert; (4) fixing ring; (6) glass; (7) ring with apertures and heater for admitting cooling air onto the glass; (8) gas outlet.

6.4. APPARATUS FOR DETERMINATION OF THE FLUORESCENCE INTENSITY DUE TO MICROSTRUCTURES IN BIOLOGICAL SPECIMENS

The luminescence of cells can be investigated, in principle, with ordinary apparatus similar to that described above by recording the total emission of a cell suspension. However, the luminescence of isolated cells and their largest microstructures is of major interest in biology. Studies of the luminescence of isolated cells which are capable of revealing fine effects but may be disturbed when the observations are carried out in a suspension can be used to relate the luminescence with the physiological state of the cell, and hence localize structural luminescence centers in the cell. Isolated cells must, of course, be investigated with the aid of a microscope. The development of special microscopes for the detection of ultraviolet fluorescence of cells became possible after the design of the "opaque-illuminator" by E. M. Brumberg. This is a special device which can be used to excite and record cell fluorescence through the same objective. The design of such microscope is a highly specialized subject and lies outside the scope of the present monograph. The requirements which the design must satisfy impose certain restrictions on the illuminating system, the diameter of the light probe, the condensers and objectives, and so on. We shall merely note that while photographic methods in the luminescence analysis of biopolymers have been made obsolescent by photoelectronics, the photographic technique continues to be used as a method of detection in the UV luminescence microscopes since it can be utilized to relate cell luminescence with its morphological properties. Electron-optical converters are also frequently employed.

Currently available UV fluorescence microscopes use photographic recording exclusively [E. M. Brumberg and I. Ya. Barskii (1961)] or additional photoelectronics for intensity measurements [Barenboim *et al.* (1962)], and a number of microspectrofluorimeters are capable of recording the UV spectra of cell fluorescence [Agroskin and Korolev (1961); Agroskin *et al.* (1964); Maksimovskii (1964); Yagunov and Prokof'ev (1966)].

The simple apparatus shown in Fig. 6.11 was developed by one of the present authors, I. Ya. Barskii, and R. I. Pinto in consultation with E. M. Brumberg. This apparatus can be used for quantitative measurements of the intensity of fluorescence due to biological microstructures. It can be used to determine the fluorescence intensity emitted by tissues, individual cells, isolated nuclei, and other cell organoids. It can also be used to localize photographically the luminescence centers in cells, and to determine the fluorescence intensity as

Fig. 6.11. Photograph of the apparatus for measuring the intensity of fluorescence of biological objects.

a function of time. The apparatus is based on an instrument developed by E. M. Brumberg and I. Ya. Barskii (1961) for UV fluorescence microscopy with photographic recording.

Block Diagram. The apparatus consists of a UV fluorescence microscope with a photometric attachment incorporating a photomultiplier (Fig. 6.12). The light source is a quartz mercury lamp 1, whose intensity is stabilized directly in the region of the cell fluorescence (see Section 5.1). The radiation from the lamp passes through a quartz collector 2, a light filter 4, a shutter 7, and is finally intercepted by an "opaque-illuminator" 14. After reflection from the beam splitter of the "opaque-illuminator," the radiation is collected by the microscope objective and is thrown onto the object under investigation in which it excites the fluorescence. The microscope objective 13 and the eyepiece 16 produce an image of the fluorescing specimen in the plane of the cathode of the photomultiplier 20.

Optical System. The "opaque-illuminator" was proposed by E. M. Brumberg in 1955. In the microscope which we are considering, the "opaque-illuminator" has two removable interference mirrors (beam splitters) (or dichroic mirrors). The first beam splitter has a reflection factor of up to 90% for ultraviolet radiation in

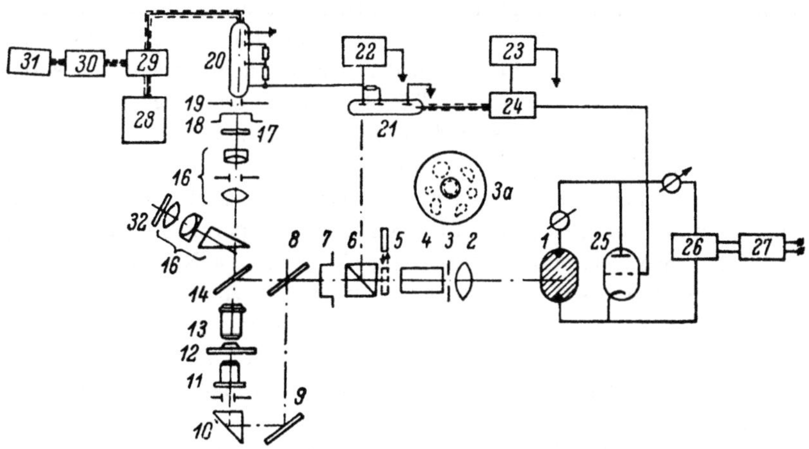

Fig. 6.12. Block diagram of the apparatus for measuring the intensity of fluorescence of biological specimens. (1) Light source; (2) collector; (3) organic glass plate with a central aperture; (3*a*) field of view with place 3 in position: all cells can be seen but only those facing the aperture in the plate luminesce; (4), (5), (17), (32) light filters; (6) tube with transparent quartz plate; (7), (18) light screens; (8)–(11) illumination system; (12) specimen (13) microscope objective; (14) opaque illuminator; (15) system for visual observation of the specimen; (16) microscope eyepiece; (19) diaphragm; (20), (21) photomultipliers; (22) photomultiplier supplies; (23) amplifier power supply; (24) stabilizer amplifier; (25) 6N5 triodes shunting the tube; (26) rectifier; (27) electromagnetic stabilizer; (28) oscillograph; (29), (30) amplifiers; (31) automatic recorder.

the range 250–280 mμ, and transmits radiation with a wavelength in excess of 300 mμ. This beam splitter is used for excitation and photoelectric recording of the ultraviolet fluorescence of cells. The second beam splitter has a high reflection coefficient in the range 360–420 mμ and good transmission at 500 mμ. It is used for the observation of cells in the light of their visible fluorescence. The specimens can also be observed by the transmitted visible radiation from the light source with the diaphragm closed or with oblique illumination. This is achieved with the aid of a system which illuminates the specimen from underneath. The illuminating part of the microscope incorporates the removable filters 4 and 5 (Fig. 6.12) defining certain regions of the spectrum which are then used to excite the fluorescence in the specimen. These filters operate in conjunction with filters 17 which are crossed with them and are placed in front of the photomultiplier cathode. The latter filters define the corresponding spectral regions in the specimen fluorescence. The gas (chlorine–bromine) filter and the UFS-1 filters can be used to isolate rays exciting the ultraviolet fluorescence of cells. These are followed by a heat filter (15-mm-thick quartz container filled with distilled water). A further light filter consisting of BS-1 and UFS-2 glasses is placed in front of the photocathode. This completely absorbs the rays which excite the fluorescence but readily transmits the UV fluorescence from the specimen. The organic glass plate 3 with a small aperture at the center is placed immediately after the field diaphragm of the illuminator. The aperture is sharply imaged by the microscope objective in the plane of the sample at the center of the field of view of the microscope. The plate enables one to see the entire field of view, and at the same time produce the fluorescence of only the particular cell which is seen in the projection of the aperture.

A diaphragm with a small aperture at its center is placed in front of the photomultiplier cathode in the plane of the cell image to suppress any scattered light.

Electronics. The photomultiplier signal is fed into a DC amplifier 29 taken from the 'Kaktus' X-ray set (Fig. 6.12). It then enters the main unit of the amplifier 30 which feeds the automatic potentiometer 31 (EPP-09). The total time for recording the signal is not more than 1 sec. When the excitation continues for less than 0.1 sec, the ENO-1 oscillograph 28 is employed for detection of the signal.

A cerium-activated fluorite plate placed in an "opaque-illuminator" 14 is used to check the stability of the entire apparatus. It serves as a standard source of fluorescence. The absorption and emission spectra of the plate are similar to those of the biological specimens under investigation, so that the standard plate can be used without replacing the light filter in the optical system The fluorescence of the cerium glass is recorded in the same way as the cell fluorescence. The instability, if any, of the apparatus as a whole can be judged from the variation of this signal. Moreover, when there is a considerable electrical and optical instability one can judge it from the ratio of the intensities emitted by the cell and the cerium plate. It is thus possible to compare the fluorescence intensities of different cells in the specimen, and of different specimens, independently of the absolute intensity of the radiation exciting the fluorescence.

The apparatus can also be used for photographing the UV fluorescence of biological objects. This is done by placing a photographic cassette under the photomultiplier tube.

6.5. APPARATUS FOR THE DETECTION OF RADIOLUMINESCENCE

The apparatus for the recording of radioluminescence must have the same general properties as in the case of photoluminescence investigations. However, the type of exciting radiation (α and β particles, γ rays, and X rays) imposes certain additional requirements. The design of the apparatus for X-ray fluorescence studies must take into account the following factors.

1. Masking effects associated with the Vavilov–Cerenkov radiation which appears in water solutions at energies in excess of 260 keV.

2. Effect of X rays on the operation of photomultipliers (direct or scattered radiation ejects electrons from the envelope, photocathode, and dynodes).

3. Masking effect due to air fluorescence, for example, the X-ray fluorescence of helium, oxygen, air, and nitrogen is in the ratio of 1:2:10:35 [Samylov *et al.* (1958)].

4. Radiolysis of the materials under investigation, which is enhanced in the presence of oxygen.

These factors are not taken into account in the existing apparatus for radioluminescence and X-ray luminescence studies of biological objects [Sapezhinskii and Emanuel' (1960); Sommermeyer *et al.* (1961)]. The authors have attempted to allow for these effects in the apparatus described below.

X-Ray Luminescence Measurements

The apparatus shown in Fig. 6.13 was designed for measuring the X-ray luminescence intensity from solutions and dry samples.

Block Diagram. The X rays from the RUM-7 tube 1 (Fig. 6.14) pass through filter 2 and enter the chamber 3. The thermally stabilized, removable bottom of this chamber contains the specimen. The fluorescence passes through the filters 5 (BS-4 and UFS-3), and is focused by a quartz lens on the photocathode 6 of the FEU-46A photomultiplier with a uviol glass window.

Excitation. The maximum X ray tube voltage for the RUM-7 tube is 60 kV, so that the quantum limit of the continuous spectrum lies at 60 keV. This means that Vavilov–Cerenkov radiation is absent. However, this voltage is close to the minimum voltage (65 kV) for which tissue radiation becomes practically uniform [Aglintsev (1957)]. The filters which were chosen experimentally (0.47 mm aluminum, and 3 mm organic glass) cut off the soft radiation which produces most of the gas luminescence.

Detection. The FEU-46A photomultiplier is placed in a metal container

Fig. 6.13. General view of the apparatus used for the study of radio luminescence (in this case the X-ray tube has been replaced by a UV source).

with a hermetically sealed textolite lid. If necessary, the container can be placed in a metal enclosure surrounded by plastic foam insulation. The metal enclosure is packed with solid carbon dioxide for cooling the photomultiplier, and silica gel is placed inside the container. An excess pressure of 0.5 atm is established to prevent the ambient air from reaching the photomultiplier and fogging its window. The photomultiplier signal is recorded by an EMU-3M electrometer.

Chamber and Specimen Container. The chamber (Fig. 6.15) has two optical windows, gas and solution inlets, and a removable thermostatted double bottom. An ebonite container is screwed into the bottom of the chamber, and the latter is lined with the same material. Ebonite is used in the part of the chamber which is exposed to X rays in order to reduce the absolute intensity of secondary emission (in comparison with the metal components) and to ensure

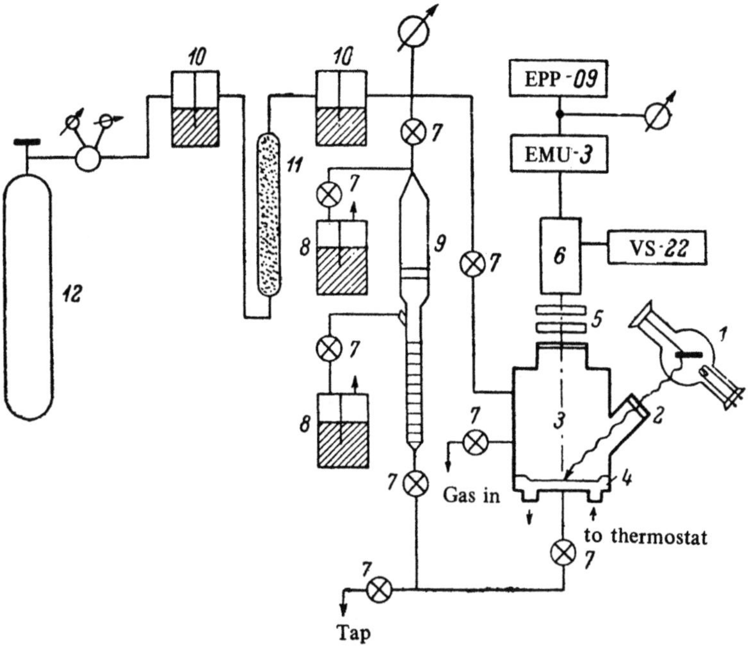

Fig. 6.14. Block diagram of the apparatus for studying X-ray luminescence. (1) X-ray tube RUM-7; (2) filters; (3) chamber; (4) specimen container; (5) light filters; (6) photomultiplier; (7) tap; (8) water trap; (9) burette with a Schott filter; (10), (11) gas purification system; (12) helium cylinder.

that its energy distribution approaches the distribution of the secondary radiation from biopolymers and cells.

Gas System. To reduce the radiolysis of solutions and gas luminescence inside the container the air is replaced by helium, which is introduced from the cylinder 12 (Fig. 6.14) through a tube into the chamber 3 and is then passed through the burette 9 which incorporates a Schott filter and is filled with the solution under investigation. By blowing helium through a water solution of tryptophan for about 20 min it is possible to reduce the oxygen content to such an extent that it is no longer detectable polarographically. The solution is then transferred to the container along the same tubes but without coming into contact with air.

Detection of X-Ray Fluorescence Spectra

As already noted (Section 3.3), X-ray fluorescence has a very low intensity. It follows that even when all measures are taken to increase the sensitivity of the apparatus (absence of Vavilov–Cerenkov radiation, replacement of air by helium suppression of scattered X rays, and photomultiplier cooling), it is not

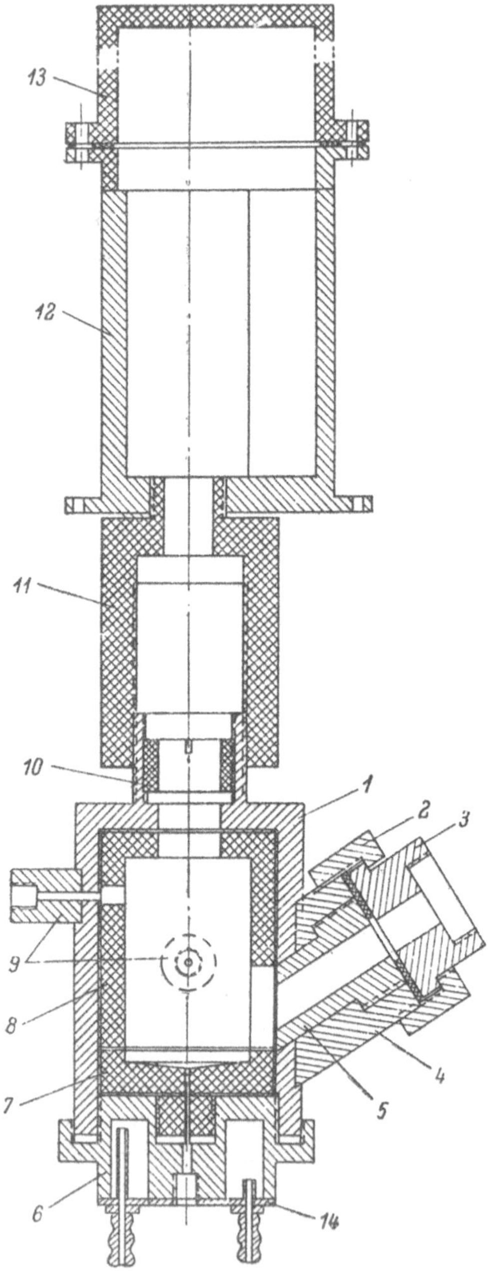

Fig. 6.15. Sectional drawing of the chamber for radioluminescence studies. (1), (4) Body; (2), (3) coupling rings for the X-ray tube; (5), (8) ebonite lining; (6) double bottom; (7) container; (9) gas inlet; (10) nut supporting light filters; (11) tube with an internal screw thread for lens; (12) photomultiplier envelope; (13) textolite cap; (14) pipes for admitting water from the thermostat.

possible to record the spectra of unfrozen solutions of proteins, nucleic acids, and their components. In most cases, the spectral composition of the X-ray fluorescence is estimated with the aid of filters placed in front of the photomultiplier [Sapezhinskii and Emanuel' (1960)]. However, in some cases, for example, in the recording of the X-ray fluorescence of crystalline tryptophan, the luminescence intensity is sufficiently high, and the spectra can be recorded with ordinary devices designed for the recording of photoluminescence spectra (with the light source replaced by the X-ray tube). In particular, the X-ray fluorescence spectra of crystalline tryptophan (Fig. 3.1) have been obtained with an apparatus (Fig. 6.16) in which the luminescence is modulated by a mechanical chopper 5. The spectrum is resolved by a double monochromator 6 (DMR-2) and is then recorded by a system consisting of the FEU-18 photomultiplier, a narrow-band amplifier with a synchronous detector, and a recorder (EPP-09).

6.6. APPARATUS FOR THE RECORDING OF ULTRAWEAK RADIATION

The detection and recording of ultraweak radiation, including the chemiluminescence of cells (dark and mitogenetic emission), became possible after the development of devices capable of recording intensities of the order of 10–100 photons/sec (Chap. 4). The radiation detector can be a cooled photomultiplier [Colli *et al.* (1954); Vasil'ev *et al.* (1959); Vladimirov and Litvin (1959*b*); Cormier (1960); Tarusov *et al.* (1962); Fontijn *et al.* (1964)], a low-noise photomultiplier [Avakyan *et al.* (1963); Agaverdiev and Tarusov (1965); Fish *et al.* (1965*b*)], or a photon counter [Troitskii *et al.* (1961)]. In principle, the chemiluminescence of cells can be recorded with any apparatus capable of responding to ultraweak radiation. This includes statistical detection [Kovaleva *et al.* (1965)], the astronomical photometer [Baum (1955); Artem'ev and Rozhinova (1963)], and UV-photon counters [Perel'man *et al.* (1959); Shelkov *et al.* (1959)].

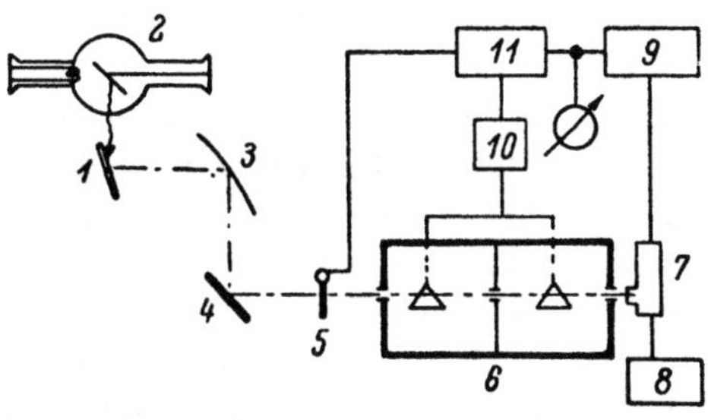

Fig. 6.16. Recording of X-ray fluorescence spectra. (1) Specimen; (2) X-ray tube; (3), (4) mirrors; (5) modulator; (6) double monochromator DMR-2; (7) FEU-18A photomultiplier; (8) voltage supplies for FEU-VS22; (9) narrow-band amplifier; (10) automatic recorder EPP-09M2 with wavelength marking; (11) phase detector.

Fig. 6.17. Photograph of the low-intensity detector. Part of the plastic foam insulation has been removed (see Fig. 6.18 for details).

When these methods are used the spectral composition of the recorded radiation can only be estimated with the aid of filters. An interesting apparatus for spectral studies of chemiluminescence has been reported by Vasil'ev (1965*c*). It incorporates an original high focal ratio monochromator [Vasil'ev *et al.* (1963)] and is capable of recording the spectra of low-intensity samples with sufficient resolution.

The apparatus developed by the present authors is described below. It is designed for the recording of ultraweak radiation in photo- and chemiluminescence studies. It can also be used to record the thermal luminescence of biopolymers and their components; this is of interest for radiobiology and photochemistry [Augenstein *et al.* (1960)]. For this purpose, the apparatus includes a semiautomatic system which can heat a frozen and irradiated object at an adjustable constant rate. Thermal luminescence curves depend on the rate of heating [Lushchik (1955)]. A photograph of the apparatus is shown in Fig. 6.17. It uses a cooled photomultiplier and an efficient light collection system.

Block Diagram. The luminescence of the sample in the container 2 (Fig. 6.18) is collected by a light pipe 5 and thrown onto the photocathode 10 of the

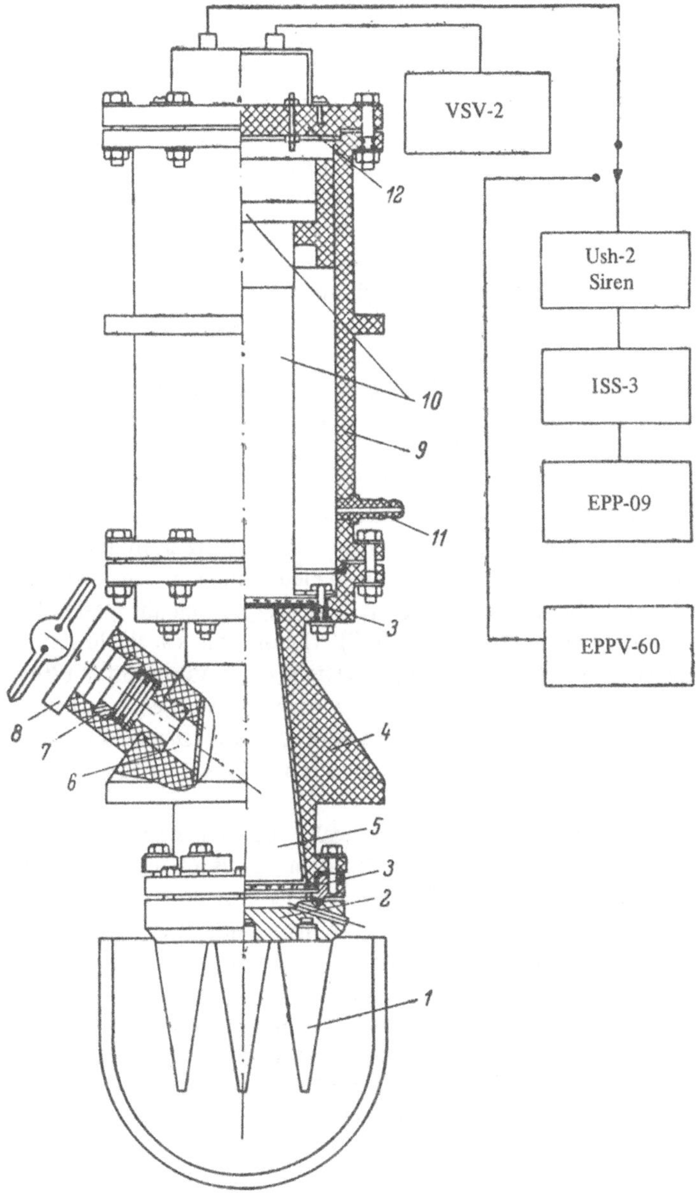

Fig. 6.18. Block diagram of the apparatus for investigating low-intensity radiation. (1) Legs; (2) sample holder; (3) lid of sample holder; (4) light-pipe case; (5) light pipe; (6) tube for exciting radiation; (7) light filters; (8) shutter; (9) photomultiplier case; (10) photomultiplier; (11) connecting pipe for liquid nitrogen vapor; (12) lid.

FEU-39 photomultiplier, which is cooled by liquid nitrogen vapor. The spectral composition of the emitted radiation is estimated by means of filters placed between the end of the light guide and the photomultiplier cathode. The fluorescence is excited through the tube 6, one end of which is covered by the shutter 8 and the other rests on the surface of the light guide. Radiation from the SVD-120 lamp, which is stabilized in intensity, enters through a lateral aperture in the light guide and is reflected from its inner surface onto the bottom of the container. The photomultiplier pulses are amplified by a broad-band amplifier (USh-2 'Siren'). They are then integrated by an ISS-3 unit and registered on the recorder EPP-09. In the case of high-intensity radiation, the signal from the photomultiplier at room temperature is recorded on a model EPPV-60, which has a high input impedance.

Light Guide. The light collection efficiency can be brought up to 70-80% by using the light guide without any careful preparation or matching of the light-guiding system. The light guide is usually made of plexiglas or glass [Matveev and Sokolov (1962); Khabakhpashev and Tseluikin (1964); Avakyan *et al.* (1965)] and is not, of course, suitable in the case of UV studies. A quartz light guide is expensive and, moreover, ordinary quartz both absorbs and fluoresces. We therefore used a hollow, thin-walled aluminum tube with a carefully polished inner surface. The light guide was designed on the basis of the data reported by Matveev and Sokolov (1962). The diameters at the lower and upper ends of the guide are 60 and 30 mm, respectively, and the height is 120 mm. The guide is hermetically sealed with quartz windows at each end, and since it is in direct contact with the photomultiplier which, in turn, is cooled by liquid nitrogen vapor, internal frosting of the windows is prevented by allowing the nitrogen vapor to enter the light guide itself.

Specimen Container. The hermetically sealed container is designed for crystalline specimens and solutions. Its diameter is 60 mm, so that a large emitting surface can be used. The container is covered with a quartz window which also seals off the lower end of the light guide. The temperature is measured by a thermocouple, and provision is made for introducing different gases and solutions. Whenever necessary, a quartz insert can be introduced to prevent the specimen from coming into contact with the metal walls. Provision is made for introducing solutions or cell suspensions. Different gases can also be introduced if necessary. The container is hermetically sealed by means of a knife-edge resting on a plastic ring. It has massive conical legs 1 which are placed in a thermostat.

★ ★ ★

Finally, we note once again that composite measurements of luminescence parameters can be successful. The fluorescence lifetime is one of the most important among these parameters. Nevertheless, we have not considered measurements of this parameter in the present monograph because of the technical difficulties involved, and partly because of the almost complete lack of published papers on such measurements on biopolymers and their monomeric chains.

One of the best examples of a useful installation for the simultaneous and comprehensive study of practically all the luminescence parameters of biological objects is that developed under the direction of Professor Tumerman at the Institute of Molecular Biology of the Academy of Sciences of the USSR [Morozov and Tumerman (1965)].

CONCLUSIONS

Biology, and especially its molecular aspects, is currently undergoing a period of rapid expansion. We shall now examine, on the basis of the material outlined in this book, what contribution luminescence methods of investigation can make to this progress and what problems face investigators who employ such methods.

1. Improvement of the methods of qualitative and quantitive analysis remains as important as ever before, and perhaps even more so in view of the fact that molecular biology makes greater demands on the purity of substances isolated by the biochemist, many of which can so far be obtained only in very small amounts.

2. Measurement of changes in the UV fluorescence of proteins is a highly sensitive method of observing their conformational transitions. Such changes may occur not only due to addition of artificial denaturing agents to proteins but also to enzyme–substrate reactions taking place *in vitro* and also when more complex systems such as muscles and certain types of cells are exposed to various influences. However, in the last two cases the application of UV fluorescence to the detection of conformational transitions requires further development.

Application of UV fluorescence may also solve a problem that itself arose in the course of luminescence investigations, i.e., the special role of aromatic residues in maintaining the protein structure and in protein–protein interactions. In this context the application of fluorescence methods to the study of the electronic structure of aromatic amino acids and the effect of solvent polarity on this structure assumes fundamental importance. In turn, the polarity of the microsurroundings of aromatic amino acid residues in the protein may also be determined with the aid of these methods.

Even where we cannot as yet say anything very definite about the molecular basis of the change in luminescence, the application of the methods of UV fluorescence to observations of the general state of normal and pathological cells is of considerable importance to medicine and cytology.

3. In the field of the luminescence of the nucleic acids, the luminescence characteristics of pure preparations of specific DNA and RNA should be measured again. Even today, however, it is clear that the study of the relative energy positions of $n\pi^*$ and $\pi\pi^*$ levels as a function of the conditions of the medium may prove of great interest, in particular in understanding the nature of mutations associated with a change in the electronic structure of nitrogenous bases. In view of the absence of luminescence from nucleic acids in neutral aqueous solutions on excitation in the normal absorption region, luminescence investigations on excitation in the region of $\sim$ 300 mμ become of considerable interest, especially as the conformational changes of DNA and its interaction with proteins (particularly histones) are reflected in the characteristics of luminescence excited in this part of the spectrum.

4. It is possible that some of the reactions occuring in living systems may betray themselves by cell luminescence which appears as a side effect, is due to oxidation processes, and lies in the visible spectrum.

However, development of the usefulness of cell luminescence requires detailed investigations whose aim should be the identification of emitting substances and the electronic states responsible for the emission. When the interaction of radicals with oxygen is one of the main and not side processes (in photochemical or radiation-chemical reactions), the chemiluminescence of biopolymers and cells may be a useful method of investigating the kinetics of various processes. An important step forward would be detection of chemiluminescence in the absence of oxygen in specific enzyme reactions, in protein-synthesizing systems, and so on.

5. The functional role of cell UV chemiluminescence (mitogenetic radiation) is still open. Stimulation of this radiation by extremely weak light fluxes from UV lamps extended considerably the possibilities of experimental investigations, but this approach is hardly the final answer.

6. Investigation of bioluminescence in cells and organisms containing specialized luciferin – luciferase systems may – quite apart from the study of this phenomenon – prove useful as a test for various influences on the cell and perhaps for studying some aspects of protein synthesis on the example of luciferase.

7. Investigations of the luminescence of biopolymers and cells are of special interest for the study of the photochemical and radiobiological reactions in which excited states undoubtedly play a definite part. In particular, study of the radioluminescence characteristics of biopolymers with regard to its dependence on the type and energy of the exciting radiation should aid in determining how much of a contribution is made by excited electronic states of macromolecules to the overall radiation poisoning.

In photochemical processes luminescence methods already make it possible to study reaction kinetics and to detect intermediate unstable products of these reactions.

8. This far from complete list of the prospects opened up in luminescence studies of cells and biopolymers indicates the need for further improvements in the methods of recording luminescence characteristics. The main problem is to increase the absolute sensitivity to weak light fluxes and to small changes in these fluxes.

Apart from traditional ways (cooling of the photomultiplier, optimized recording circuits, reduction of light losses, and so on), it is useful to study the physical processes occurring in the photomultipliers for a reasonably broad-based approach to the task of reducing the photomultiplier noise level.

Equally important problems exist in the field of manufacturing high-transmission monochromators, cells from nonluminescent materials, and powerful and stable UV sources. In connection with the growing interest in the study of the intrinsic luminescence of complex enzyme–substrate, antigen–antibody, and other systems, there is now a need for spectrometers for the recording of UV luminescence with allowance for the specific features of biological investigations.

Two main attitudes can be discerned among investigators with regard to luminescence methods. The optimists regard these methods as omnipotent; they believe that the high sensitivity, effective lack of inertia, and the possibility of working on living objects ensure infinite prospects. On the other hand, the skeptics hold that the observed photoluminescence changes cannot be unequivocally associated with the causative processes.

It is our hope that the data presented in this book will help the reader to define for himself the true limits of the application of luminescence to biology, and will promote the solution of problems in the investigation of which luminescence methods can play a decisive part.

While this book was in preparation a number of publications appeared on closely related topics. The most important of these publications are included in the bibliography.

The main laws of luminescence are the subject of an extensive chapter in the book by Seliger and McElroy (1965), which devotes considerable attention to the luminescent properties of heteroatomic molecules. Photoluminescence of proteins is discussed in detail in books by Vladimirov (1965) and Konev (1965). The problems of chemiluminescence are the subject of books by Vladimirov (1966), Shlyapintokh *et al.* (1966), Seliger and McElroy (1965), Vasil'ev's review (1966), and symposia on free-radical processes [Shlyapintokh (1966)] and chemiluminescence [Wyman (1965)]. The modern ideas about the physics of the processes and photoelectronic instruments, and also the characteristics of new types of photomultipliers, are given in a book by Soboleva *et al.* (1965).

BIBLIOGRAPHY

Agaverdiev, A.Sh. and Tarusov, B.N. (1965). *Biofizika* **10**(2):351–352.

Agaverdiev, A.Sh., Doskoch, Ya.E., and Tarusov, B.N.(1965*a*). *Biofizika* **10**(5):832–836.

Agaverdiev, A.Sh., Doskoch, Ya. E., and Tarusov, B.N. (1965*b*). *Dokl. Akad. Nauk SSSR* **163**(4):991–993.

Aglintsev, K.K. (1957). *The Dosimetry of Ionizing Radiations* (in Russian), Moscow.

Agroskin, L.S. (1957). *Biofizika* **2**(4): 518–519.

Agroskin, L.S., Korolev, N.V., Kulaev, I.S., Meisel', M.N., and Pomoshchnikova, N.A. (1960). *Dokl. Akad. Nauk SSSR* **131**(6):1440–1443.

Agroskin, L.S. and Barskii, I.Ya. (1961). *Dokl. Akad. Nauk SSSR* **139**(4): 987–990.

Agroskin, L.S. and Korolev, N.V. (1961). *Biofizika* **6**(4):478–485.

Agroskin, L.S., Korolev, N.V., Kulaev, I.S., and Pomoshchnikova, N.A. (1961*a*). *Dokl. Akad. Nauk SSSR* **136**(1):226–229.

Agroskin, L.S., Korolev, N.V., Kulaev, I.S., and Pomoshchnikova, N.A. (1961*b*). *Dokl. Akad. Nauk SSSR* **131**(6):1440–1443.

Agroskin, L.S. and Pomoshchnikova, N.A. (1962). *Biofizika* **7**(3):292–297.

Agroskin, L.S. and Korolev, N.V. (1963). *Pribory i Tekhn. Eksperim.* No.4, pp. 135–137.

Agroskin, L.S., Predtechenskii, Yu.B., Barskii, I.Ya., and Shudel', M.S. (1964). In: *Electron and Fluorescent Microscopy of Cells* (in Russian), Moscow, pp.152–157.

Ainsworth, S. and Winter, E. (1964). *Appl. Optics,* **3**(3):371–383.

Aksentsev, S.L., Olenev, V.I., and Vladimirov, Yu.A. (1965). "Spectra of Action of Fluorescence Induced by Visible Light in Aromatic Acids Irradiated at 77°K by Ultraviolet Radiation." *Biofizika* **10**(5):877–878.

Aleksandrov, O.V., Grofren, M.V., Keilina, T.A., and Tveryankina, R.I. (1965). *Zh. Prikl. Spektroskopii Akad. Nauk Belorussk. SSR* **2**(5):471–475.

Alentsev, M.N. and Vinokurov, L.A. (1951). *Izv. Akad. Nauk SSSR, Ser. Fiz.* **15** (6):725–729.

Allabutaev, K.A., Vasil'ev, R.F., Vichutinskii, A.A., and Rusina, I.F. (1965). In: *Bioluminescence* (in Russian), Moscow, pp. 8–18.

Arevshatyan, A.A. (1966). "Quantitative Evaluation of Weak Biological Luminescence." *Izv. Akad. Nauk Arm. SSR* **19**(1):73–77.

Arnold, W. and Sherwood, H.K. (1957). *Proc. Nat. Acad. Sci. USA* **43**(1):105–114.

Artem'ev, V.V. and Rozhinova, I.P. (1963). *Izv. Krymsk. Astrofiz. Observ.* **30**:297–307.

Artem'ev, V.V. (1964). *Radiotekhn. i Elektron.* **9**(4):756–757.

Audubert, R. (1938). *Angew. Chem.* **51**(11):153–163.

Audubert, R. (1939). *Trans. Faraday Soc.* **35**(213):197–204.

Augenstein, L.G., Carter, J.G., Nelson, D.R., and Jockey, H.P. (1960). *Radiation Res. Suppl.* **2**:19.

Augenstein, L.G., Carter, J.G., Nelson, D.R., and Jockey, H.P. (1961). In: *Free Radicals in Biological Systems,* New York, pp.149–156.

Augenstein, L. (1962). *Advan. Enzymol.* **24**:359–413.

Augenstein, L. (1963). In: *Progress in Biophysics (Molecular Biology)* Vol. 13, New York, pp. 1–58.

Augenstein, L. and Nag-Chaudhuri, J. (1964). *Nature* **203** (4950):1145–1147.

Augenstein, L., Brustad, T., and Mason, R. (1964*a*). In: *Advances in Radiation Biology* Vol. 1, New York, pp. 227–266.

Augenstein, L., Carter, J., Nag-Chaudhuri, J., Nelson, D., and Veargerd, E. (1964*b*). In: *Physical Processes in Radiation Biology,* New York, pp. 73–89.

Avakyan, Ts.M., Adzhyan, N.S., and Apayan, R.R. (1963). *Biofizika* **8**(4):385–386.

Avakyan, Ts.M., Tarusov, B.N., and Adgoyan, S.N. (1965). In: *Bioluminescence* (in Russian), Moscow, pp. 60–63.

Bacq, Z.M. and Alexander, P. (1961). *Fundamentals of Radiobiology,* Oxford.

Baev, A.A. (1965). In: *Biosynthesis of Albumin and Nucleic Acids* (in Russian), Moscow, pp. 126–171.

Bakhshiev, N.G. (1961). *Opt. i Spektroskopiya* **30**(6):717–726.

Bakhshiev, N.G. (1964). *Opt. i Spektroskopiya* **16**(5):821–832.

Baldwin, G.C. and Friedman, S.I. (1965). *Rev. Sci. Instr.* **36**(1):16–18.

Barenboim, G.M., Barskii, I.Ya., and Pinto, R.I. (1961). *Radiobiologiya* **1**(6):843–850.

Barenboim, G.M. (1962). *Biofizika* **7**(2):227–232.

Barenboim G.M., Barskii, I.Ya., Brumberg, E.M., and Pinto, R.I. (1962). *Biofizika* **7**(3):351–356.

Barenboim, G.M. (1963*a*). *Interaction of Oxygen with the Molecules of Aromatic Aminoacids, Albumins, and Nucleic Acids in the Excited Triplet State* (in Russian), Leningrad.

Barenboim, G.M. (1963*b*). *Interaction of Oxygen with the Molecules of Aromatic Aminoacids, Albumins, and Nucleic Acids in the Excited Singlet State* (in Russian), Leningrad.

Barenboim, G.M. (1963*c*). *Biofizika* **8**(2):154–163.

Barenboim, G.M. and Domanskii, A.N. (1963). *Biofizika* **8**(3):321–329.

Barenboim, G.M., Pinto, R.I., and Pravdina, K.I. (1963*a*). *Radiobiologiya* **3**(1):8–11.

Barenboim, G.M, Domanskii, A.N., and Pravdina, K.I. (1963*b*). In: *Fifteenth Conference on Spectroscopy, Abstracts* (in Russian), Minsk.

Barenboim, G.M., Domanskii, A.N., and Rozanov, Yu.M. (1964*a*). In: *Thirteenth Conference on Luminescence, Abstracts* (in Russian), Moscow, p. 8.

Barenboim,G.M., Domanskii, A.N., and Solomatin, V.F. (1964*b*). In: *Thirteenth Conference on Luminescence, Abstracts* (in Russian), Moscow, pp. 8–9.

Barenboim, G.M. (1966). In: *Handbook of Cytology Vol. 2* (in Russian), Moscow, pp. 126–137.

Barenboim, G.M. and Mochalkin, A.I. (1966). *Radiobiologiya* **6**(5):671–674.

Barenboim, G.M., Domanskii, A.N., and Solomatin, V.F. (1966). *Pribory i Tekhn. Eksperim.* No. 2, pp. 131–133.

Barenboim, G.M. and Kudasheva, A.M. (1967). "Use of an External Electrostatic Shield to Reduce Noise in a Photoelectronic Multiplier." *Materials of the Scientific Conference of the Institute of Cytology of the Academy of Sciences of the USSR* (in Russian), Leningrad, pp. 11–12.

Barenboim, G.M., Domanskii, A.N., and Solomatin, V.F. (1967*a*). In: *Industrial Chemical Reagents. Technical and Economical Information Series* (in Russian), Moscow, pp. 254-263.

Barenboim, G.M, Danilov, V.I., Kruglyak, Yu.A., and Umetskaya, V.N. (1967*b*). "Quantum

Mechanical Calculation of the Electronic Structure of Aromatic Amino Acids and Certain Precursors of Their Synthesis." *Materials of the Scientific Conference of the Institute of Cytology of the Academy of Sciences of the USSR* (in Russian), Leningrad, pp. 10–11.

Barenboim, G.M, Domanskii, A.N., Rozanov, Yu.M., and Turoverov, K.K. (1967*c*) in: *Industrial Chemical Reagents. Technical and Economical Information Series*, No.8(14) (in Russian), Moscow, pp. 264–274.

Barenboim, G.M., Verkhovtseva, M.I., Domanskii, A.N., Ruban, E.L., Sobota, A., and Suvorov, N.N. (1967*d*) "Study of the Exogenous Accumulation of L-tryptophan in the Presence of Its Precursors in Yeasts *Candida utilis* by a Luminesence Method." *Izv. Akad. Nauk SSSR, Ser. Biol.* No. 5 pp. 766–777.

Barenboim, G.M., Bozhkov, V.M., Vinokurova, T.I., Vorob'ev, V.M., Gavrilova, L.P., Domanskii, A.N., Pluchennichak, A., Spirin, A.S., and Turoverov, K.K. (1967*e*). "Fluorescence of the Ribosomes." *Materials of the Scientific Conference of the Institute of Cytology of the Academy of Sciences of the USSR* (in Russian), Leningrad, p.8.

Barenboim, G.M., Sokolenko, V.A., and Turoverov, K.K. (1968). "Investigation of Temperature Influences on Proteins by a Fluorescence Method." *Tsitologiya* **12**(5) (in press).

Barskii, I.Ya. and Brumberg, E.M. (1958) *Biokhimiya* **23**(5):791–792.

Barskii, I.Ya. (1959) *Biokhimiya* **24**(5):823–825.

Barskii, I.Ya., Brumberg, E.M., Bukhman, M.P., Vasilevskaya, E.I., and Pluzhnikova, G.F. (1959). *Botan. Zh.* **44**(5):639–642.

Barskii, I.Ya., Brumberg, E.M., and Kondrat'eva, T.M. (1961). *Biofizika* **6**(5):606–609.

Barskii, I.Ya. and Brumberg, I.E. (1962). *Tsitologiya* **4**(3):328–330.

Barskii, I.Ya., Brumberg, E.M., and Brumberg, V.A. (1962). *Dokl. Akad. Nauk SSSR* **147**(2):474–476.

Barskii, I.Ya. (1964). *Development of Photographic and Photoelectric Methods and Apparatus for the UV Fluorescent Microscopy for Biological Objects, Abstract of Dissertations* (in Russian). Leningrad.

Barskii, I.Ya. and Yakubenas, V.A. (1964). In: *Electron and Fluorescent Microscopy of Cells* (in Russian), Moscow, pp. 164–168.

Barskii, I.Ya., Rozanov, Yu.M., and Shiffers, L.A. (1967). "A Microfluorimeter for Investigating the Polarized Fluorescence of Biological Specimens." *Tsitologiya* **9**(8):1026–1030.

Barskii, I.Ya., Rozanov, Yu.M., Chernogryadskaya, N.A., Shiffers, L.A., and Shudel', M.S. (1968). "Polarized Fluorescence of Microstructures of Biological Specimens." *Izv. Akad. Nauk SSSR, Ser. Fiz.* (in press).

Barskii, V., Zelenin, A.V., and Lyapunova, E.A. (1962). In: *Eleventh Conference of Luminescence, Abstracts* (in Russian), Moscow, p.6.

Basu, S. and Lox, L. (1962). *J. Chim. Phys.* **59**(10):1031–1032.

Basu, S. and Greist, I.H. (1963). *J. Chim. Phys.* **60**(3):407–411.

Basu, S. and Lox, L. (1963). *Biochim. Biophys. Acta* **76**(1):131–135.

Baum, W.A. (1955). *Astron. J.* Vol. 60,No. 1(1224), pp. 25–28.

Bay, Z. and Papp, G. (1964). *Trans. IEEE PGNS* **11**(3):160–167.

Beaven, G.H., Holiday, E.R., and Johnson, E.A. (1955). In: *The Nucleic Acids* Vol. 1, New York, pp. 493–553.

Beaven, G.H. (1962). In: *Advances in Spectroscopy*, Vol 2, New York, pp. 341–428.

Becher, O. (1957). *Biol. Zentr.* **76**(2):141–147.

Belikova, T.P. (1959). *Opt. i Spektroskopiya* **6**(1):117–118.

Berezin, I.V., Denisov, E.T., and Emanuel', N.M. (1962). *Oxidation of Cyclohexane*

(in Russian), Moscow.
Bersohn, R. and Isenberg, I. (1963). *Biochem. Biophys. Res. Commun.* **13**(3):205–208.
Bersohn, R. and Isenberg, I. (1964). *J. Chem. Phys.* **40**(11):3175–3180.
Birks, J.B. (1964). *The Theory and Practice of Scintillation Counting,* London.
Birks, J.B., Braga, C.L., and Lumb, M.D. (1964). *Brit. J. Appl. Phys.* **15**(4):399–404.
Biryuzova, V.I. and Meisel', M.N. (1964). In: *Molecular Biology* (in Russian), Moscow, pp. 316–323.
Bither, B. and McElroy, W. (1957). *Arch. Biochem. Biophys.* **72**(2):358–368.
Blumenfel'd, L.A.. (1959). *Biofizika* **4**(5):515–520.
Bobrovich, V.P. and Konev, S.V. (1964). *Dokl. Akad. Nauk SSSR* **155**(1):197–200.
Bobrovich, V.P. and Konev, S.V. (1965). *Dokl. Akad. Nauk Belorussk. SSR* **9**(2):118–121.
Bobrovich, V.P., Chernitskii, E.A., and Konev, S.V. (1965). *Pribory i Tekhn. Eksperim.* No.2, pp. 187–188.
Borisevich, N.A. (1967). *Excited States of Complex Molecules in the Gas Phase* (in Russian), Minsk.
Borisov, A.Yu. (1960). *Opt. i Spektroskopiya* **9**(1):115–116.
Borisov, A.Yu. and Mokhova, E.N. (1964). *Pribory i Tekhn. Eksperim.* No.2, pp. 145–147.
Borresen, H.C. (1963*a*). *Acta Chem. Scand.* **17**(4):921–929.
Borresen, H.C. (1963*b*). *Acta Chem. Scand.* **17**(8):2359–2360.
Bowman, R.L., Caulfield, P.A., and Udenfriend, S. (1955). *Science* **122**(2157):32–33.
Brand, L., Everse, Y., and Kaplan, N.O. (1962). *Biochemistry* **1**(3):423–432.
Bresler, S.E. (1963). *Introduction to Molecular Biology* (in Russian), Moscow.
Bresler, V.M., Kudryavtseva, M.V., Pil'shchik, E.M. Chernogryadskaya, N.A., and Shudel', M.S. (1964*a*). In: *Electron and Fluorescent Microscopy of Cells* (in Russian), Moscow, pp. 93–95.
Bresler, V.M., Chernogradskaya, N.A., Pil'shchik, E.M., Kurdyavtseva, M.V., Shudel', M.S., Rozanov, Yu.M., Astashina, T.N., and Veselkina, M.P. (1964*b*). In: *Thirteenth Conference on Luminescence, Abstracts* (in Russian), Moscow, p.17.
Bresler, V.M. (1965). In: *Physical and Chemical Self-Regulation in Cells, Abstracts* (in Russian), Moscow, p. 17.
Bresler, V.M. and Vorob'ev, V.I. (1965). In: *Electron and Fluorescent Microscopy of Cells* (in Russian), Moscow, pp. 124–130.
Bresler, V.M, Brumberg, E.M., Kudryavtseva, M.V., Pil'shchik, E.M., Chernogryadskaya, N.A., and Shudel', M.S. (1965*a*). *Byull. Eksperim. Biol. i Med.* **59**(5):89–92.
Bresler, V.M., Chernogryadskaya, N.A., Pil'shchik, E.M., Shudel', M.S., Kudryavtseva, M.V., Astashina, T.P., Rozanov, Yu.M., and Selivanova, G.V. (1965*b*). In: *Physical and Chemical Self-Regulation in Cells, Abstracts* (in Russian), Moscow, pp. 9–10.
Bresler, V.M., Chernogryadskaya, N.A., Pil'shchik, E.M., Shudel', M.S., Kudryavtseva, M.V., Selivanova, G.V., Rozanov, Yu.M., Veselkina, M.P., and Astashina, T.P. (1966). "Quantitative Study of Proliferating Normal and Tumoral Liver Cells of Rats and Mice. 1. Cytophotometric Investigations of the Regulation of Cell Metabolism in Living Isolated Cells of the Regenerating Rat Liver." *Tsitologiya* **8**(1):59–69.
Bresler, V.M., Chernogryadskaya, N.A., Pil'shchik, E.M., Pyl'dvere, E.I., Shudel', M.S., Kudryavtseva, M.V., Selivanova, G.V., Rozanov, Yu.M., Veselkina, M.N., Vorob'ev, V.I., and Astashina, T.P. (1968). *Normal and Pathological Cytology of the Liver* (in Russian), Leningrad (in press).
Broomhead, J.M., Gelboin, H.V., Luborsky, S.W., Richards, H.H., and Singer, M.F., (1962). *Biochim. Biophys. Acta* **61**(3):354–367.
Bruckner, B.H. (1964). *Radiation Res.* **22**(1):171–172.
Brugger, J.E. and Franck, I. (1958). *Rev. Biochem. Biophys.* **75**(2):465–496.
Brumberg, E.M. (1955). *Zh. Obshch. Biol.* **16**(3):222–237.

Brumberg, E.M. (1956). *Zh. Obshch. Biol.* **17**(6):401–412.
Brumberg, E.M., Meisel', M.N., Barskii, I.Ya., and Bukhman, M.P. (1958). *Zh. Obshch. Biol.* **19**(2):99–107.
Brumberg, E.M. and Barskii, I.Ya. (1960). *Tsitologiya* **2**(3):318–324.
Brumberg, E.M., Barskii, I.Ya., and Shudel', M.S. (1960*a*). *Tsitologiya* **2**(5):589–591.
Brumberg, E.M., Barskii, I.Ya., Kondrat'eva T.M., Chernogryadskaya, N.A., and Shudel', M.S. (1960*b*). *Dokl. Akad. Nauk SSSR* **135**(6):1521–1524.
Brumberg, E.M. and Barskii, I.Ya. (1961). *Zh. Obshch. Biol.* **22**(6):459–466.
Brumberg, E.M., Barskii, I.Ya., and Kondrat'eva, T.M. (1961*a*). *Biofizika* **6**(5):605-609.
Brumberg, E.M., Barskii, I.Ya., Kondrat'eva, T.M., and Chernogryadskaya, N.A. (1961*b*). *Biofizika* **6**(1):114–118.
Brumberg, E.M., Meisel', M.N., Barskii, I.Ya., Zelenin, A.V., and Lyapunova, E.A. (1961*c*). *Dokl. Akad. Nauk SSSR* **141**(3):723–725.
Brumberg, E.M., Barskii, I.Ya., Chernogryadskaya, N.A., and Shudel', M.S. (1963*a*). *Dokl. Akad. Nauk SSSR* **150**(6):1356–1358.
Brumberg, E.M., Barskii, I.Ya., Chernogryadskaya, M.A., and Shudel', M.S. (1963*b*). *Izv. Akad. Nauk SSSR, Ser. Biol.* No. 1, pp. 87–90.
Brumberg, E.M. and Brumberg, I.E.(1964). *Biofizika* **9**(6):748–750.
Brumberg, E.M., Chernogryadskaya, N.A., and Shudel', M.S. (1964). In: *Electron and Fluorescent Microscopy of Cells* (in Russian), Moscow, pp. 74–81.
Brumberg, E.M. and Brumberg, I.E. (1965). "Some Possible Regulators of Cell Growth in the Organism." *Dokl. Akad. Nauk SSSR* **165**(5):1171–1174.
Brumberg, E.M. and Chernogryadskaya, N.A. (1965). "Fluorescence Microscopy." In: *Handbook of Cytology, Vol. 1* (in Russian), Moscow, pp. 64–71.
Brumberg, I.E. (1964). *Biofizika* **9**(3):382–383.
Brumberg, E.M., Barskii, I.Ya., and Papayan, G.V. (1967). "A Two-wave Microfluorimeter." *Opt. Mekh. Prom.* No. 9, p. 62.
Brumberg, I.E. and Brumberg, E.M. (1964*a*). *Biofizika* **9**(2):237–238.
Brumberg, I.E. and Brumberg, E.M. (1964*b*). *Biofizika* **9**(4):502–503.
Brumberg, I.E. and Dunaeva, S.E. (1964). *Tsitologiya* **6**(6):766–768.
Buck, J.B. (1961). in: *Light and Life*, Baltimore, pp. 754–770.
Buck, J.B. and Case, I.F. (1961). In: *Light and Life*, Baltimore, pp.754–770.
Buck, W.L. (1960). *Trans. IRE PGNS* **7**(2–3):11–16.
Bull, H.B. (1947). *Physical Biochemistry*, New York.
Burns, C.H., Nevill, A.D., and Whipple, G.H. (1962) *Science* **135**(3502):430–432.
Burns, C.H. and Whipple, G.H. (1964). *Radiation Res.* **22**(1):173.
Burns, C.H., Jacobson, A.P., and Whipple, G.H. (1965). *Radiation Res.* **24**(3):494–502.
Burshtein, E.A. (1961). *Biofizika* **6**(6):753–763.
Burshtein, E.A. (1964). *Application of Spectroscopic Methods to Studies of the Functional States of Cells* (in Russian), Moscow.
Burshtein, E.A. and Vladimirov, Yu.A. (1964). *Biofizika* **10**(1):48–55.
Burshtein, E.A. and Suslova, T.B. (1964). *Biofizika* **9**(2):184–193.
Burshtein, E.A. and Vladimirov, Yu.A. (1964). *Biofizika* **9**(2):184–193.
Burshtein, E., Zimina, G.M., and Vladimirov, Yu.A. (1966). "Use of the Luminescence Method to Determine the State of Aromatic Amino Acids (Tyrosine and Tryptophan) in Proteins and Enzyme Systems." *Zh. Prikl. Spektroskopii Akad. Nauk Belorussk. SSR* **4**(2):142–146.
Burton, M. (1960). *Radiation Res. Suppl.* **2**:119.
Carter, G., MacRae, R.A., Nelson, D.R., Burkhoff, R.D., and Arakawa, E.T. (1964). *Appl. Opt.* **3**(1):75–78.
Carter, J.G., Nelson, D.R., and Augenstein, L.G. (1965). *Arch. Biochem. Biophys.*

111(2):270–282.
Chance, B. (1940). *Electronics* **13**(2):24–25.
Charlesby, A. (1960). *Atomic Radiation and Polymers,* New York.
Chase, A.M. (1964). In: *Photophysiology,* New York, pp. 389–421.
Chechik, N.O., Fainshtein, S.M., and Lifshits, G.M. (1957). *Electron Multipliers* (in Russian), Moscow.
Chernitskii, E.A., Konev, S.V., and Bobrovich, V.P. (1963). *Dokl. Akad. Nauk Belorussk. SSR* **7**(9):628–632.
Chernitskii, E.A. and Konev, S.V. (1965). *Zh. Prikl. Spektroskopii Akad. Nauk Belorussk. SSR* **2**(3):261–266.
Chernogryadskaya, N.A. and Shudel', M.S. (1962*a*). *Dokl. Akad. Nauk SSSR* **145** (4):917–919.
Chernogryadskaya, N.A. and Shudel', M.S. (1962*b*). In: *Second Scientific Conference of the Institute of Cytology, Abstracts* (in Russian), Leningrad, pp. 82–83.
Chernogryadskaya, N.A., Buze, E.G., and Astashina, T.N. (1964*a*). In: *Thirteenth Conference in Luminescence, Abstracts* (in Russian), Moscow, p. 114.
Chernogryadskaya, N.A., Pil'shchik, E.M., Shudel', M.S., Kudryavtseva, M.V., and Astashina, T.P. (1964*b*). *Dokl. Akad. Nauk SSSR* **156**(1):174–176.
Chernogryadskaya, N.A., Pil'shchik, E.M., Shudel', M.S., Kudryavtseva, M.V., and Astashina, T.P. (1964*c*). In: *Electron and Fluorescent Microscopy of Cells* (in Russian), Moscow, pp. 82–87.
Chernogryadskaya, N.A., Brumberg, E.M., Bresler, V.M., Pil'shchik, E.M., Shudel', M.S., Kudryavtseva, M.V., and Astashina, T.N. (1964*d*). In: *Second International Congress of Histo- and Cytochemistry,* Frankfurt-on-Main, p. 217.
Chernogryadskaya, N.A., Bresler, V.M., Pil'shchik, E.M., Kudryavtseva, M.V., and Shudel', M.S. (1965). In: *Physical and Chemical Self-Regulation in Cells, Abstracts* (in Russian), Moscow, pp. 42–43.
Chizhikova, Z.A. (1961). *Tr. Fiz. Inst. Akad. Nauk SSSR* **15**:178–229.
Chumakova, R.I. (1963*a*). *Izv. Sibirsk. Otd. Akad. Nauk SSSR, Ser. Biol. Med. Nauk* No. 2 pp. 62–67.
Chumakova, R.I. (1963*b*). *Interrelation between the Luminescent Reaction of Bacteria and Some Aspects of Their Metabolism, Abstracts* (in Russian), Krasnoyarsk.
Chumakova, R.I. and Egorova, A.L. (1964*a*). *Mikrobiologiya* **33**(3):423–427.
Chumakova, R.I. and Egorova, A.L. (1964*b*). *Mikrobiologiya* **33**(4): 639–642.
Clarke, W.G. (1963). *Nature* **198**(4887):1244–1246.
Cochran, W. (1951). *Acta Cryst.* **4**(2):81–92.
Colli, L. and Facchini, U. (1954). *Nuovo Cimento* **12**(1):150.
Colli, L., Facchini, U., and Rossi, A. (1954). *Nuovo Cimento* **11**(3):255–261.
Colli, L., Facchini, U., Guidotti, G., Dugnani Lonati, R., Orsenigo, M., and Sommariva, O. (1955). *Experientia* **11**(12):479–481.
Cormier, M.J. (1960). *Biochem. Biophys. Acta* **42**(2):333–337.
Cormier, M.J. (1961). In: *Light and Life,* Baltimore, p. 274–293.
Cormier, M.J. and Dure, L.S. (1963). *J. Biol. Chem.* **238**(2):785–789.
Cormier, M.J. and Totter, I.R. (1964). In: *Ann. Rev. Biochem.* **33**:431–458.
Cornog, J.L. and Adams, W.R. (1963). *Biochim. Biophys. Acta* **66**(3):356–365.
Cowgill, R.W. (1963*a*). *Arch. Biochem. Biophys.* **100**(1):36–44.
Cowgill, R.W. (1963*b*). *Biochim. Biophys. Acta* **75**(2):272–273.
Cowgill, R.W. (1964*a*). *Arch. Biochem. Biophys.* **104**(2):84–92.
Cowgill, R.W. (1964*b*). *Biochem. Biophys. Res. Commun.* **16**(4):332–335.
Cowgill, R.W. (1965*a*). *Biochim. Biophys. Acta* **94**(1):74–80.

Cowgill, R.W. (1965*b*). *Biochim. Biophys. Acta* **94**(1):81–88.

Dalinenko, N.K. and Razumovskaya, A.P. (1966). "A Photoelectronic Multiplier Sensitive in the Ultraviolet Region of the Spectrum." *Pribory i Tekhn. Eksperim.* No. 3, pp. 212–213.

Dantsing, N.M. (ed.) (1966). *Ultraviolet Radiation* (in Russian), Moscow.

Debye, P. and Edwards, I.O. (1952). *Science* **116**(3006):143–144.

De Luca, M., Wirtz, G.W., and McElroy, W.D. (1964). *Biochemistry* **3**(7):935–939.

De Luca, M. and McElroy, W.D. (1965). *Biochem. Biophys. Res. Commun.* 18(5–6): 836–842.

Dexter, D.L. (1953). *J. Chem. Phys.* **21**(5):836–850.

Dimartin, M., Lochet, R., Rybak, B., and Rousset, A. (1957). *Compt. Rend.* **244**(24): 2905–2909.

Domanskii, A.N., Pluchennichak, A., Rozanov, Yu.M., and Turoverov, K.K. (1966). In: *Proceedings of the First Scientific Conference of Young Specialists, Institute of Cytology of the Academy of Sciences of the USSR* (in Russian), Leningrad, No. 16.

Domanskii, A.N., Barenboim, G.M., Verkhovtseva, M.I., and Sobota, A. (1967). "Spectral Investigations of Certain Indole Derivatives and Precursors of Tryptophan Biosynthesis." *Materials of the Scientific Conference of the Institute of Cytology of the Academy of Sciences of the USSR* (in Russian), Leningrad, pp. 39–40.

Donovan, L.G. (1963). *Biochim. Biophys. Acta* **78**(3):474–491.

Doty, P. and Gratzer, W.B. (1962). In: *Polyamino Acids, Polypeptides, and Proteins*, Madison, pp. 111–118.

Douzou, P., Francq, J., Hanss, M., and Ptak, M. (1961). *J. Chim. Phys.* **58**(10):926–933.

Douzou, P. and Francq, J. (1962). *J. Chim. Phys.* **59**(6):578–583.

Dubinin, N.P. (1961). *Problems of Radiation Genetics* (in Russian), Moscow.

Duggan, D.E., Bowman, R.L., Brodie, B.B., and Udenfriend, S. (1957). *Arch. Biochem. Biophys.* **68**(1):1–14.

Dunaevskaya, N.V., Soshin, L.D., and Yushkevich, G.F. (1966). "The Optimum Spectrometric System of the Photoelectronic Multiplier with a Louvered Dynode System." *Pribory i Tekhn. Eksperim.* No. 1, pp. 198–199.

Dupay, F., Lochet, R., and Rousset, A. (1964). *Compt. Rend.* **258**(17):4223–4227.

Dure, L.S. and Cormier, M.J. (1963*a*). *J. Biol. Chem.* **238**(2):790–793.

Dure, L.S. and Cormier, M.J. (1963*b*). *Biochem. Biophys. Res. Commun.* **11**(6):489–495.

Dure, L.S. and Cormier, M.J. (1964). *J. Biol. Chem.* **239**(7):2351–2359.

Dzhafarova, A.P., Kol's, O.R., and Tarusov, B.N. (1965). "Biochemiluminescence of Muscle and Liver Homogenates of Warm-Blooded Animals, Preserved by Deep-Freezing." In: *Problems of Internal Reception and the Physiological Activity of Substances* (in Russian), Baku, pp. 77–84.

Eberhardt, E.H. (1964). *Trans. IEEE PGNS* **11**(3):48–55.

Edelhoch, H. and Steiner, R.F. (1962). *Biochim. Biophys. Acta* **60**(2):365–372.

Edelhoch, H. and Steiner, R.F. (1963). *J. Biol. Chem.* **238**(3):931–938.

Edelhoch, H. Brand, L., and Wilchek, M. (1963). *Israel J. Chem.* **1**(3*a*):216–217.

Edelhoch, H. and Steiner, R.F. (1964). In: *Electronic Aspects of Biochemistry*, New York, pp. 7–22.

Edelhoch, H., Frattali, V., and Steiner, R.F. (1965). *J. Biol. Chem.* **240**(1):122–127.

Eidus, L.Kh. (1964). In: *Principles of Radiation Biology* (in Russian), Moscow, pp. 11–42.

Eisenberg, L., Rosen, P., and Edelman, G.M. (1962). *Rev. Sci. Instr.* **33**(12):1435–1440.

Eley, D.D. (1962). In: *Horizons in Biochemistry*, New York, pp. 341–380.

El'yashevich, M.A. (1962). *Atomic and Molecular Spectroscopy* (in Russian), Moscow.

Emanuel, N.M., Kruglyakova, K.E., Zhizhina, G.P., Vichutinskii, A.A., and Vasil'ev, R.F. (1965). In: *Bioluminescence* (in Russian), Moscow, pp. 119–121.

Eremeev, V.F. (1961). *Byul. Eksperim. Biol. i Med.* **51**(4):102–106.

Ermolaev, V.L. and Terenin, A.N. (1960). *Usp. Fiz. Nauk* **71**(1):137–141.
Ermolaev, V.L. and Sveshnikova, E.B. (1962). *Izv. Akad. Nauk SSSR, Ser. Fiz.* **26**(1):29–31.
Ermolaev, V.L. (1963). *Usp. Fiz. Nauk* **80**(1):3–40.
Ermolaev, V.L. and Sveshnikova, E.B. (1963). *Dokl. Akad. Nauk SSSR* **149**(6):1295–1298.
Ermolaev, V.L. and Sveshnikova, E.B. (1964). *Opt. i Spektroskopiya* **16**(4):587–593.
Fano, U. (1953*a*). *Nucleonics* **11**(8):8–12.
Fano, U. (1953*b*). In: *Symposium on Radiobiology*, Oxford, pp. 87–96.
Fano, U. (1961). In: *Comparative Effects of Radiation*, New York, pp. 14–21.
Fasman, G., Norland, K., and Pesce, A. (1964). In: *Quantum Aspects of Polypeptides and Polynucleotides*, New York, pp. 325–331.
Feitelson, J. (1964). *J. Phys. Chem.* **68**(2):391–397.
Feofilov, P.P. (1948). *Usp. Fiz. Nauk* **36**(4):417–455.
Feofilov, P.P. (1960). *Physical Basis of Polarized Emission*, New York.
Fish, A.M., Chumakova, R.I., and Salanskii, N.M. (1965*a*). In: *Controlled Biosynthesis and Biophysics of Populations* (in Russian), Krasnoyarsk, p. 153.
Fish, A.M., Salanskii, N.M., and Chumakova, R.I. (1965*b*). In: *Bioluminescence* (in Russian), Moscow, pp. 177–180.
Fleishman, D.G. and Shakhidzhanyan, L.G. (1959). *Pribory i Tekhn. Eksperim.* No. 1, pp. 135–136.
Fleishman, D.G. (1962). *Pribory i Tekhn. Eksperim.* No. 5, pp. 98–102.
Fontijn, A., Meyer, C.B., and Shiff, H.I. (1964). *J. Chem. Phys.* **40**(1):64–70.
Förster, T. (1948). *Ann. Physik* **2**:55–75.
Förster, T. (1959). *Discussions Faraday Soc.* **27**:7–17.
Franck, J. and Sponer, H. (1956). *J. Chem. Phys.* **25**(1):172.
Franklin, A.R., Holloway, W.W., and McMahon, D.H. (1965). *Rev. Sci. Instr.* **36**(2):232.
Frattali, V., Steiner, R.F., and Edelhoch, H. (1965). *J. Biol. Chem.* **240**(1):112–121.
Freed, F. and Salmre, W. (1958). *Science* **128**(3335):1341–1342.
Fujimori, E. (1960). *Biochim. Biophys. Acta* **40**(2):251–256.
Gachkovskii, V.F. (1960). *Dokl. Akad. Nauk SSSR* **133**(6):1358–1360.
Galanin, M.D. and Frank, I.M. (1951). *Zh. Eksperim. i Tekhn. Fiz.* **21**(2):114–120.
Galanin, M.D. (1960). *Tr. Fiz. Inst. Akad. Nauk SSSR* **12**:3–53.
Galanin, M.D. and Chizhikova, Z.A. (1961). *Opt. i Spektroskopiya* **11**(2):271–272.
Gally, J.A. and Edelman, G.M. (1962). *Biochim. Biophys. Acta* **60**(3):499–509.
Gally, J.A. and Edelman, G.M. (1964). In: *Biopolym. Symp. 1*, New York, pp. 367–381.
Galston, A.W. (1950). *Science* **111**(2893):619–624.
Gandy, H.W. and Weller, J.F. (1964). *Rev. Sci. Instr.* **35**(3):413–414.
Gasanov, R.A. and Mamedov, T.G. (1963). *Nauchn. Dokl. Vysshei Shkoly, Biol. Nauki* No. 3, pp. 88–92.
Gasanov, R.A., Mamedov, T.G., and Tarusov, B.N. (1963*a*). *Dokl. Akad. Nauk SSSR* **150**(4):913–915.
Gasanov, R.A., Mamedov, T.G., and Tarusov, B.N. (1963*b*). *Dokl. Akad. Nauk SSSR* **153**(4): 947–949.
Gasanov, R.A. (1965). "Mechanisms of Biochemiluminescence of Plants." *Izv. Azerbaidzhanskoi Akad. Nauk SSSR, Ser. Biol. Nauk*, No. 3, pp. 9–15.
Gauroy, P. and Grive, J. (1964). *Nature (Paris)*, No. 3347, pp. 98–104.
Gavrilenko, L.M., Lebed', G.G., and Nazarov, D.A. (1962). *Pribory i Tekhn. Eksperim.* No. 3, p. 193.
Gawronski, E. (1961). *Ann. Univ. Mariae Curie-Sklodovska, Sect. C.* **14**(19):407–433.
Gemiel, C.L. (1956). *Anal. Chem.* **28**(6):1061–1064.

Gerstein, G.F., Van Vunakis, H., and Levine, L. *Biochemistry* **2**(5):964–971.

Gibson, Q.H. and Hastings, J.W. (1962). *Biochem. J.* **83**(2):368–377.

Giese, A. (1957). *Cell Physiology*, Philadelphia.

Gill, J.E. (1963). *Dissertation Abstr.* **24**(1):354.

Gill, J.E. and Weissbluth, M. (1964). In: *Biopolym. Symp. 1*, New York, pp. 433–439.

Gitel'zon, I.I. and Chumakova, R.I. (1965). In: *Controlled Biosynthesis and Biophysics of Populations* (in Russian), Krasnoyarsk, p. 152.

Gitel'zon, I.I., Chumakova, R.P., and Fish, M.A. (1965*a*). "A Bathyphotometer and the Bioluminescence of the Waters of the Pacific and Indian Oceans." In: *Bionics* (in Russian), Moscow, pp. 314–323.

Gitel'zon, I.I., Chumakova, R.I., and Fish, A.M. (1965*b*). *Biofizika* **10**(1):100–104.

Gitel'zon, I.I., Fish, A.M., and Chumakova, R.I, (1965*c*). In: *Controlled Biosynthesis and Biophysics of Populations* (in Russian), Krasnoyarsk, p. 153.

Gitel'zon, I.I., Fish, A.M., and Chumakova, R.I. (1965*d*). "Apparatus for Investigating the Dynamic Characteristics of Metabolism during Continuous Culturing of Microorganisms." *Mikrobiologiya* **34**(6):1086–1091.

Gladchenko, L.F., Kostko, M.Ya., Pikulik, L.G., and Sevchenko, A.N. (1965). "Duration of the Excited State of Ultraviolet Fluorescence of Aromatic Amino Acids." *Dokl. Akad. Nauk Belorussk. SSR* **9**(10):647–650.

Glukhovskii, B.M. and Osherovich, A.L. (1967). "Parameters of Photomultipliers with a Small Working Area of the Photocathode." *Prikl. Spektroskopii Akad. Nauk Belorussk.* **6**(1):10–15.

Gol'dshtein, I.I., Roshupkin, D.I., and Skulachev, V.P. (1966). "Low-Temperature Luminescence Spectra of Adenine and Its Derivatives." *Biokhimiya* **31**(5):814–818.

Golland, M.I. (1961). *Apparatus for Luminescence Analysis* (in Russian), Moscow.

Golubeva, G.I., Rozenkrants, S.A., and Garina, L.P. (1965). *Opt. Mekh. Prom.* No. 4, pp. 1–6.

Gray, L.H. (1955). *Actions Chimiques et Biologiques de Radiations* Vol. 1., Masson.

Green, A. and McElroy, W.D. (1956). *Biochim. Biophys. Acta* **20**(1):170–176.

Green, D.E. and Fleischer, S. (1962). In: *Horizons of Biochemistry*, New York, pp. 381–420.

Green, D.E. (1964). In: *Molecular Biology* (in Russian), Moscow, pp. 260–271.

Grinberg, A.A. (1920). *Zh. Russk. Fiz. Khim. Obshch.* **52**(4–5–6):151–185.

Groves, J.R.V. (1965). *J. Sci. Instr.* **42**(6):430–431.

Gurvich, A.A. (1960). *Byul. Eksperim. Biol. i Med.* **49**(5):482–485.

Gurvich, A.A. (1961). *Byul. Eksperim. Biol. i Med.* **51**(5):68–73.

Gurvich, A.A. (1964). *Byul. Mosk. Obshchestva Ispytatelei Prirody, Otd. Biol.* **69**(6):145.

Gurvich, A.A. (1965). *Biofizika* **10**(4):619–624.

Gurvich, A.A. and Eremeev, V.F. (1965). In: *Bioluminescence* (in Russian), Moscow, pp. 135–142.

Gurvich, A.A., Eremeev, V.F., and Karabchievskii, Yu.A. (1965) *Nature* **206**(4979):20–22.

Gurvich, A.A. (1966). "Spectra of the Mitogenetic Radiation of the Muscles as an Index of the Dynamic Organization of the Sarcoplasm." *Byul. Eksperim. Biol. i Med.* **62**(8):53–56.

Gurvich, A.A. and Eremeev, V.F. (1966). "Mitogenetic Radiation of the Cardiac Muscle in Animals." *Byul. Eksperim. Biol. i Med.* **61**(6):56–58.

Gurvich, A.A., Eremeev, V.F., and Sobieva, Z.P. (1966). "Mitogenetic Radiation of a Partially Deafferentated Cardiac Muscle." *Byul. Eksperim. Biol. i Med.* **62**(7):55–59.

Gurvich, A.G. and Gurvich, L.D. (1934). *Mitogenic Radiation* (in Russian), Leningrad.

Gurvich, A.G. and Gurvich, L.D. (1947). In: *Collection of Papers on Mitogenesis and the Theory of the Biological Field* (in Russian), Moscow, pp. 5–20, 120–130.

Gurvich, A.G. and Gurvich, L.D. (1948). *Introduction to Mitogenesis* (in Russian), Moscow.

Gurvich, A.G. (1959). *Die mitogenetische Strahlung,* Jena.

Gusar, D.N., Starinkin, Yu.A., and Belova, N.G. (1965). In: *Physical-Mathematical Methods in Biology and Medicine* (in Russian), Novosibirsk, p. 46.

Hanter, F.E. (1962). In: *Fifth International Biochemical Congress* (in Russian) Moscow, pp. 301–320.

Hart, E.J. and Platzman, R.L. (1961). In: *Mechanisms in Radiobiology,* New York, pp. 93–258.

Hartmann, W. and Bernhard, F. (1957). *Fotovervielfacher und ihre Anwendung in der Kernphysik.*

Harvey, E.N. (1940). *Living Light,* New Jersey.

Harvey, E.N. (1952). *Bioluminescence,* New York.

Hastings, J.W., Riley, W.H., and Massa, J. (1965). *J. Biol. Chem.* **240**(3):1473–1781.

Haugen, G.R. and Marcus, R.J. (1964). *Appl. Opt.* **3**(9):1049–1056.

Hayman, L.G., Schwarz, R.M., and Schluter, R.A. (1964). *Rev. Sci. Instr.* **35**(3):393–406.

Hercules, D.M. (1964). *Science* **145**(3634):808–809.

Herskovits, T.T. and Laskowski, M. (1962). *J. Biol. Chem.* **237**(11):3418–3422.

Hine, G.Y. and Brownell, G.L. (eds.) (1956). *Radiation Dosimetry,* New York.

Hirata, Y., Shimomura, O., and Eguchi, S. (1959). *Tetrahedron Letters,* No. 5, pp. 4–9.

Hollander, A. (1960). *Quart. Rev. Biol.* **35**(3):246–247.

Howard-Flanders, P. (1957). *Nature* **180**(4596):1191–1192.

Howard-Flanders, P. and Jockey, P. (1960). *Radiation Res.* **13**(3):466–478.

Isenberg, I. and Szent-Györgyi, A. (1958). *Proc. Acad. Sci., USA* **44**(6):519–521.

Isenberg, I. and Bersohn, R. (1964). In: *Electronic Aspects of Biochemistry,* New York, pp. 3–5.

Ismailova, S. Kh., Kozlov, Yu.P. and Gasanov, (1965). In: *Physical-Chemical Self-Regulation in Cells* (in Russian), Moscow, pp. 19–20.

Ivanov, I.I. and Petrusevich, Yu.M. (1965). *Nauchn. Dokl. Vysshei Shkoly, Biol. Nauki* No. 3, pp. 81–82.

Johnson, F.H. (ed.) (1955). *Luminescence in Biological System,* Washington.

Johnson, F.H., Eyring, H., and Chang, J.J. (1959). *Discussions Faraday Soc.* **27**:191–199.

Johnson, F.H., Sie, E.H.-C, and Haneda, Y. (1961). In: *Light and Life,* Baltimore, pp. 206–218.

Johnson, F.H., Shimomura, O., Saiga, Y., Gershman, L.C., Reynolds, G.T., and Waters, I.R. (1962). *J. Cellular and Comp. Physiol.* **60**(1):85–103.

Kallman, H., Kramer, B., and Sucov, E. (1955). *J. Chem. Phys.* **23**:1043.

Kallman, H. and Furst, M. (1958). In: *Scintillation Counting,* New York, pp. 3–23.

Kaminskii, M.G. (1959). *Opt. i Spektroskopiya* **6**(1):103–106.

Kanda. Y. and Shimada, R. (1961). *Spectrochim. Acta* **17**(3):279–285.

Kanda, Y., Shimada, R., and Takenoshita, Y. (1963). *Spectrochim. Acta* **19**(7):1249–1260.

Karakashian, M.W. and Hastings, J.W. (1962). *Proc. Nat. Acad. Sci. USA* **48**(12):2130–2137.

Karakashian, M.W. and Hastings, J.W. (1963). *J. Gen. Physiol.* **47**(1):1–12.

Karyakin, A.V. (1961). *Usp. Khim.* **30**(8):1050–1068.

Kasha, M. (1950). *Discussions Faraday Soc.* **9**:14–19.

Kasha, M. (1961). In: *Light and Life,* Baltimore, pp. 31–64.

Kasha, M. (1960). *Radiation Res. Suppl.* **2**:243–275.

Kasha, M. (1964). In: *Physical Processes in Radiation Biology,* New York, pp. 90–92.

Katibnikov, M.A. and Konev, S.V. (1961). *Pribory i Tekhn. Eksperim.* No. 4, pp. 166–169.
Katibnikov, M.A. and Konev, S.V. (1962). *Biofizika,* 7(2):150–153.
Kaznacheev, V.P., Kuznetsov, P.G., and Subbotin, M.Ya. (1964). *Mechanisms of Sclerotic Processes and Scarring* (in Russian), Novosibirsk, pp. 49–61.
Kaznacheev, V.P. (1965). In: *Physical-Mathematical Methods in Biology and Medicine* (in Russian), Novosibirsk, pp. 38–41.
Kaznacheev, V.P., Kulikova, L.A., and Starinkin, Yu. A. (1965*a*). In: *Physical-Mathematical Methods in Biology and Medicine* (in Russian), Novosibirsk, p. 44.
Kaznacheev, V.P., Artem'ev, V.V., Goldobin, A.S., Gus'kov, A.N., Kamenskaya, V.V., and Kulikova, L.A. (1965*b*) In: *Physical-Mathematical Methods in Biology and Medicine* (in Russian), Novosibirsk, pp. 34–35.
Kaznacheev, V.P., Kamenskaya, V.V., Kulikova, L.A., Starinkin, Yu.A., Kulikov, V.Yu., Nabiulin, M.S., and Gusar, D.N. (1965*c*) In: *Physical-Mathematical Methods in Biology and Medicine* (in Russian), Novosibirsk, pp. 36–37.
Kaznacheev, V.P., Ivanov, G.K., Kamenskaya, V.V., Kazanina, S.S., Kuznetsov, P.G., Nabiulin, M.S., Subbotin, M.Ya., and Yakobson, G.S. (1965*a*) In: *Controlled Biosythesis and the Biophysics of Populations* (in Russian), Krasnoyarsk, p.152.
Kaznacheev, V.P., Artem'ev, V.V., Goldobin, A.S., Gus'kov, L.N., Kamenskaya, V.V., Kulikova, L.A., and Zenin, A.P. (1967). "Recording the Optical Emission of Biological Tissues." *Izv. Sibirsk. Otd. Akad. Nauk SSSR. Ser. Biol. Med. Nauk,* 12(3):114–118.
Kembrovskii, G.S., Bobrovich, V.P., and Konev, S.V. (1966). "Low-Temperature Luminescence Spectra of Indole." *Zh. Prikl. Spektroskopii Akad. Nauk Belorussk. SSR* 5(5):695–698
Khabakhpashev, A.G. and Tseluikin, V.A. (1964). *Pribory i Tekhn. Eksperim.* No. 1, pp. 202–203.
Khan-Magometova, Sh.D., Gutkina A.B., Meisel', M.N., Agroskin, L.S., and Korolev, N.V. (1960). *Biofizika* 5(4):446–449.
Khlebnikov, N.S., Melamid, A.E., and Timoshenkov, Yu.A. (1961). *Pribory i Tekhn. Eksperim.* No. 3. pp.129–131.
Khlebnikov, N.S., Melamid, A.E., and Kovaleva, T.A. (1962). *Radiotekh. i Elektron.* No. 3. pp. 516–521.
Khrushchev, N.G. (1963). *Arkh. Anat. Gistol. i Embriol.* 46(2):39–45.
Kihlman, B.A. (1958). *Exptl. Cell Res.* 14(3):639–642.
Kilin, S.F. and Rozman, I.M. (1963). *Opt. i Spektroskopiya* 15(4):494–499.
Kilin, S.F., Mikhelashvili, M.S., and Rozman, I.M. (1964). *Opt. i Spektroskopiya* 16(4):663–673.
Kimball, R.F. (1955). In: *Radiation Biology Vol. 2,* New York, pp. 285–314.
Kislyak, G.M. (1958). *Izv. Krymsk. Ped. Inst.* 29:358–369.
Klipson, N.A. and Mamedov, T.G. (1965). "Chemiluminescence in the Oxidation of the Substrate of the Peroxidase – Hydrogen Peroxide System." *Byul. Mosk. Obshchestva Ispytatelei Prirody Otd. Biol.* 70(1):171–172.
Klipson, N.A., Mamedov, T.G., and Tarusov, B.N. (1965). In: *Bioluminescence* (in Russian), Moscow, pp. 107–111.
Kohnlein, N. and Muller, A. (1962). *Phys. Med. Biol.* 6(3):599–604.
Kondrat'eva, T.M., Safronova, V.G., Barskii, I.Ya., and Brumberg, E.M. (1964) In: *Electron and Fluorescence Microscopy of Cells* (in Russian), Moscow, pp. 108–113.
Konev, S.V. and Katibnikov, M.A.(1961*a*). *Dokl. Akad. Nauk SSSR* 136(2):472–465.
Konev, S.V. and Katibnikov, M.A. (1961*b*). *Biofizika* 6(6):638–644.
Konev, S.V. and Lyskova, T.I. (1961). *Biofizika* 10(4):694–695.

Konev, S.V., Katibnikov, M.A., and Petrova, M.A. (1961*a*). *Biofizika* 6(3):375.
Konev, S.V., Troitskii, N.A., and Katibnikov, M.A. (1961*b*). In: *Fifteenth International Biochemical Congress Vol. 2.* (in Russian), Minsk, pp. 35–36.
Konev, S.V., Lyskova, T.I., and Bobrovich, V.P. (1963*a*). *Biofizika* 8(4):433–440.
Konev, S.V., Lyskova, T.I., and Saloshko, P.N. (1963*b*). *Biofizika* 8(2):260–268.
Konev, S.V. (1964). *Vestsi Akad. Nauk Belorusk. SSR* No. 3. pp. 59–68.
Konev, S.V. and Chernitskii, E.A. (1964). *Biofizika* 9(4):520–522.
Konev, S.V. and Lyskova, T.I. (1964). In: *Biological Effect of Ultraviolet Radiation* (in Russian), Moscow, p. 83.
Konev, S.V., Katibnikov, M.A., and Lyskova, T.I. (1964). *Biofizika* 9(1):124–127.
Konev, S.V. (1965). *Electronically Excited States of Biopolymers* (in Russian), Minsk.
Konev, S.V. and Bobrovich, V.P. (1965). *Biofizika* 10(5):813–816.
Konev, S.V. and Chernitskii, E.A. (1965). *Dokl. Akad. Nauk Belorusk. SSR* 9(5):328–330.
Konev, S.V. and Katibnikov, M.A. (1965). In: *Bioluminescence* (in Russian), Moscow, pp. 69–74.
Konev, S.V. and Lyskova, T.I. (1965*a*). "Effects of Ultraweak Intensities of UV Rays on Cell Division and Glycolysis." *Biofizika* 10(6):1000–1002.
Konev, S.V. and Lyskova, T.I. (1965*b*). "Effect of the Acceleration of Glycolysis in Yeast Cells by Ultraviolet Light of Mitogenetic Intensity." *Dokl. Akad. Nauk Belorussk. SSR* 9(3):190–193.
Konev, S.V. and Lyskova T.I. (1965*c*). *Biofizika* 9(4):520–522.
Konev, S.V., Bobrovich, V.P., and Chernitškii, E.A. (1965*a*). *Biofizika* 10(1):42–47.
Konev, S.V., Bobrovich, V.P., and Lyskova, T.I. (1965*b*). "Correlation between the Parameters of Ultraviolet Luminescence of Cells and Their Tissue and Species Classification." In: *Biophysics of the Cell* (in Russian), Moscow, pp. 98–104.
Konev, S.V., Bobrovich, V.P., and Chernitskii, E.A. (1965*c*). "Possibilities and Mechanisms of Energy Migration in Proteins." *Dokl. Akad. Nauk SSSR* 165(4):937–939.
Konev, S.V., Lyskova, T.I., and Nisenbaum, G.D. (1965*d*). In: *Controlled Biosynthesis and Biophysics of Population* (in Russian), Krasnoyarsk, pp. 150–151.
Konev, S.V. and Bobrovich, V.P. (1966). "Phosphorescence Spectra of Proteins at Various Temperatures." *Dokl. Akad. Nauk Belorussk. SSR* 10:786–788.
Konev, S.V. and Lyskova, T.I. (1966). "On the Existence of Nonchemical (Physical) Mechanisms of Interaction between Cells at a Distance." *Izv. Akad. Nauk Belorussk SSR Ser. Biyal.* No. 2, pp. 85–88.
Konev, S.V. and Volotovskii, I.D. (1966). "Investigation of the Role of Singlet and Triplet Excited States of Tryptophan in the Photoactivation of Trypsin." *Biofizika* 11(5): 791–795.
Konev, S.V., Bobrovich, V.P., and Kembrovskii, G.S. (1966*a*). "Effectiveness of Intertryptophan Energy Migration in Proteins." *Biofizika* 11(1):174–175.
Konev, S.V., Chernitskii, E.A., and Muraveinik, V.R. (1966*b*). "Temperature Dependence of the Degree of Polarization of the Fluorescence of Proteins." *Dokl. Akad. Nauk Belorussk. SSR* 10:341.
Konev, S.V., Kostko, M.Ya., Pikulik, L.G., and Chernitskii, E.A. (1966*c*). "Duration of the Excited State of Fluorescence of Proteins." *Biofizika* 11(6):965–969.
Konev, S.V., Lyskova, T.I., and Nisenbaum, G.D. (1966*d*). "Ultraweak Bioluminescence in the Ultraviolet Region of the Spectrum and Its Biological Role." *Biofizika* 11(2):361–363.
Konev, S.V., Kostko, M.Ya., Pikulik, L.G., and Volotovskii, I.D. (1966*e*). "Possible Causes of Differences in the Duration of the Excited State of Fluorescence of Protein Solutions." *Dokl. Akad. Nauk Belorussk. SSR* 10:500.
Konev, S.V. (1967). *Investigation of the Luminescence of Aromatic Amino Acids and Pro-*

teins in Solution and in Cells (Doctoral Dissertation, in Russian), Moscow.
Konstantinova-Shlezinger, M.A. (ed.) (1961). *Luminescence Analysis* (in Russian). Moscow.
Korten, E. (1964). *Chemiker Ztg.* 88:385–388.
Kovaleva, T.A., Melamid, A.E., Pisarevskii, A.N., Reznikov, I.V., and Shushkevich, S.S. (1965). *Pribory i Tekhn. Eksperim.* 2:150–154.
Kovaleva, T.A., Kuptsova, G.Z., and Melamid, A.E. (1966*a*). "Calculation of the Threshold Sensitivity of Photoelectronic Multipliers." *Radiotekhn. i Elektron.* **11**(3):568–570.
Kovaleva, T.A., Melamid, A.E., Pertsev, A.N., and Pisarevskii, A.N. (1966*b*). "Noise of Photoelectronic Multipliers." *Pribory i Tekhn. Eksperim.* No. 5 pp. 5–12.
Krasnoviskii, A.A. (1963). In: *Bioluminescence* (in Russian), Moscow, pp. 16–17.
Ksanfômaliti, L.V. (1962). *Pribory i Tekhn. Eksperim.* No. 2, pp. 121–123.
Kudryavtseva, M.V. (1967). *Cytophysiological Investigations of Liver Parenchyma Cells of White Rats in Late Embryogenesis.* (Candidate's Dissertation, in Russian), Leningrad.
Kulin, E.T. (1960). *Dokl. Akad. Nauk SSSR* **4**(2):78–81.
Kulin, E.T. (1965). In: *Bioluminescence* (in Russian), Moscow, pp. 196–197.
Kuppermann, A. and Raff, L.M. (1964). In: *Physical Processes in Radiation Biology,* New York, pp. 161–182.
Kuwabara, S., Kreiss, P., and Cormier, M. (1964). *Federation Proc.* **23**(2) Part 1, p. 163.
Kuwabara, S., Cormier, M.Y., Dure, L.S., Kreiss, P., and Pfuderer, P. (1965) *Proc. Nat. Acad. Sci. USA* **53**(4):822–828.
Kuzin, A.M. (1958). In: *Radiobiology* (in Russian), Moscow, pp. 3–13.
Kuznetsov, A.N., and Kukushkin, A.K. (1965). In: *Controlled Biosynthesis and Biophysics of Populations* (in Russian), Krasnoyarsk, p. 144.
Lamola, A.A., Leermakers, P.A., and Byers, G.W. (1965). *J. Am. Chem. Soc.* **87**(11):2322–2332.
Lea, D.E. (1959). *Action of Radiations on Living Cells,* Cambridge.
Lehninger, A.L. (1962*a*). In: *Horizons of Biochemistry,* New York, pp. 421–435.
Lehninger, A.L. (1962*b*). In: *Proceedings Fifth International Biochemical Congress* (in Russian), Moscow, pp. 253–274.
Lehninger, A.L. (1964). *The Mitochondrion. Molecular Basis of Structure and Function,* New York.
Lehrer, S.S. and Fasman, G. (1964). *Biopolymers* **2**(2):199–203.
Leiteizen, L.G. and Glukhovskii, B.M. (1964). *Izv. Akad. Nauk SSSR Ser. Fiz.* **28**(1):115–117.
Levshin, V.L. (1951). *Photoluminescence of Liquids and Solids* (in Russian), Moscow.
Liebson, S.H., Bishop, M.E., and Elliott, J.O. (1950). *Phys. Rev.* **80**:907–999.
Likhtenshtein, G.I. and Purmal', A.P. (1965). In: *Controlled Biosynthesis and Biophysics of Populations* (in Russian), Krasnoyarsk, p. 186.
Likhtenshtein, G.I. and Purmal', A.P. (1966). "Chemiluminescence Accompanying the Catalase Process." *Biofizika* **11**(2):245–248.
Lippincott, S.C., Schoof, C.Y., and Lawton, A.H. (1964). *Federation Proc.* **23**(2): Part 1, p. 213.
Lipsky, S. (1964). In: *Physical Processes in Radiation Biology,* New York, pp. 215–232.
Litvin, F.F., Vladimirov, Yu.A., and Krasnovskii, A.A. (1960). *Usp. Fiz. Nauk* **71**(1):149–155.
Litvin, F.F. and Personov, R.N. (1961). *Dokl. Akad. Nauk SSSR* **136**(4):798–800.
Lohmann, W., Fowler, C.F., and Perkins, W.H. (1963). *Acta Radiol.* **1**(3):183–189.
Lohmann, W., Fowler, C.F., and Perkins W.H. (1964). *Acta Radiol.* **2**(1):17–23.

Longin, P. (1959). *Compt. Rend.* **248**(13):1971–1973.

Longworth, J.W. (1961). *Biochem. J.* **81**(1):24p–25p.

Longworth, J.W. (1962). *Biochem. J.* **84**(3):104p–105p.

Lushchik, Ch.B. (1955). *Studies of Capture Centers in Alkali-Halide Crystal Phosphors* (in Russian), Tartu.

Lyalin, G.N. and Kobyshev, G.I. (1963). *Dokl. Akad. Nauk SSSR* **148**(4):1053–1059.

Maksimovskii, L.F. (1964). *Tsitologiya* **6**(2):255–256.

Mamedov, T.G. (1965). *Physical-Chemical Self-Regulation in Cells* (in Russian), Moscow, p. 32.

Manteifel, V.N. and Meisel, M.N. (1962). *Radiobiologiya* **2**(1):101–104.

Marshak, I.S. (1962). *Pribory i Tekhn. Eksperim.* No.3. pp. 5–22.

Marshak, I.S. (1963). *Pulsed Sources of Light* (in Russian), Moscow.

Mashanskii, V.F. (1965). In: *Handbook of Cytology Vol. 1.* (in Russian), Moscow, pp. 200–218.

Mason, R. (1959). *Discussions Faraday Soc.* **27**:129–133.

Masterson, T.S., Mason, A.D., and Brown, W.L. (1964). *Am. J. Med. Electron.* **3**(4):283–285.

Mataga, N., Torihasi, J., and Esumi, K. (1964). *Theoret. Chim. Acta* **2**(2):158–167.

Matheson, R.M. (1964). *Trans IEEE PGNS* **11**(3):64–71.

Matveev, B.B. and Sokolov, A.D. (1962). *Photomultipliers in Sintilation Counters* (in Russian), Moscow.

Matveev, V.V., Minaeva, E.E., and Sokolov, A.D. (1962). *Pribory i Tekhn. Eksperim.* No. 1. pp. 144–148.

McClure, D.S. (1951). *J. Chem. Phys.* **19**(6):670–675.

McClure, D.S. (1959). *Solid State Phys.* **8**(1):1–47.

McElroy, W.D. and Strehler, B.L. (1954). *Bacteriol. Rev.* **18**(2):177–194.

McElroy, W.D. (1960). *Federation Proc.* **19**(4): Part 1, pp. 941–948.

McElroy, W.D. (1961). in: *Bacteria, Vol. 2,* New York, pp. 479–508.

McElroy, W.D. and Seliger, H.H. (1961). In: *Light and Life,* Baltimore, pp. 219–257.

McElroy, W.D. (1962). In: *Enzymes, Vol. 6,* New York, pp. 433–442.

McElroy, W.D. and Seliger, H.H. (1962*a*). *Sci. Am.* **207**(6):76–89.

McElroy, W.D. and Seliger, H.H. (1962*b*). In: *Horizons in Biochemistry,* New York, pp. 91–101.

McElroy, W.D. and Seliger, H.H. (1962*c*). *Federation Proc.* **21**(6):1006–1012.

McElroy, W.D. and Seliger, H.H. (1962*d*). In: *Proceedings of the Fifth International Biochemical Congress, Symposium 3, Evolutionary Chemistry,* Moscow, pp. 161–171.

McElroy, W.D. (1963*a*). In: *Methods in Enzymology Vol. 6,* New York, pp. 775–777.

McElroy, W.D. (1963*b*). In: *Methods in Enzymology Vol. 6,* New York, pp. 445–448.

McElroy, W.D. and Seliger, H.H. (1963). In: *Advances in Enzymology Vol. 25,* New York, pp. 119–166.

McGlynn, S.P., Smith, F.J., and Cilento, G. (1964). *Photochem. Photobiol.* **3**(4):269–294.

McLaren, A.D. and Shugar, D. (1964). In: *Photochemistry of Proteins and Nucleic Acids,* Oxford, pp. 29–87.

Meier, A. and Zetts, E. (1952). *Ultraviolet Radiations* (in Russian), Moscow.

Meisel', M.N., Brumberg, E.M., Kondrat'eva, T.M., and Barskii, I.Ya. (1961). in: *The Initial Effect of Ionizing Radiations on Cells,* New York, pp. 107–129.

Meisel', M.N., Gutkina, A.V., Zelenin, A.V., Ognevetskaya, N.E., Vorotnitskaya, N.E., Lyapunova, E.A., and Barskii, V.E. (1962). In: *Eighth International Cancer Congress Vol. 4* (in Russian), pp. 71–73.

Meisel', M.N. and Manteifel', V.M. (1963). *Izv. Akad. Nauk SSSR Ser. Biol.* No. 6, pp. 871–879.

Mekshenkov, M.I. and Andreitsev, A.P. (1960). *Biofizika* 5(5):616–620.
Mekshenkov, M.I. and Andreitsev, A.P. (1961). *Biofizika* 6(5):615–619.
Melhuish, W.H. and Murashige, R.H. (1962). *Rev. Sci. Inst.* 33(11):1213–1215.
Mkrtchyan, R.G., Zakaryan, A.E., Petrosyan, E.P., Kovalenko, T.A., and Garibyan, D.V. (1965). In: *Physical-Chemical Self-Regulation in Cells* (in Russian), Moscow, p. 34.
Mochalkin, A.I. (1965). In: *Collection of Papers on Agronomic Physics Vol. 12* (in Russian), Leningrad.
Moiseeva, M.N. (1960*a*). *Ukr. Bot. Zh.* 17(4):29–35.
Moiseeva, M.N. (1960*b*). *Mitogenetic Rays and Mitogenetic Methods* (in Russian), Kiev.
Morgulis, N.D. and Dyatlovitskaya, (1940). *Zh. Tekhn. Fiz.* 10(2):657–660.
Morozov, Yu.V. and Tumerman, L.A. (1965). In: *Controlled Biosynthesis and Biophysics of Populations* (in Russian), Krasnoyarsk, p. 156.
Morrison, P. (1952). In: *Symposium on Radiobiology*, Oxford, pp. 53–71.
Nag-Chaudhuri, J. and Augenstein, L. (1964). *Quantum Aspects of Polypeptides and Polynucleotides,* New York, pp. 441–452.
Naray, Zs. (1956). *J. Sci. Inst.* 33(12):476–478.
Nasonov, D.N. and Aleksandrov, V.Ya. (1940). *Reaction of Living Matter to External Effects* (in Russian), Moscow.
Nasonov, D.N. (1959). *Local Protoplasm Reaction and Propagating Excitation* (in Russian), Moscow.
Neifakh, S.A. (1964). In: *Molecular Biology* (in Russian), Moscow, pp. 273–280.
Nelipa, N.F. (1960). *Introduction to the Theory of Multiple Scattering* (in Russian), Moscow.
Neporent, B.S. and Bakhshiev, N.G. (1960). *Opt. i Spektroskopiya,* 8(6):777–786.
Neporent, B.S. (ed.) (1966). *Elementary Photoprocesses in Molecules.* Moscow.
Nicol, J.A.C. (1960). *The Biology of Marine Animals,* New York.
Nicol, J.A.C. (1962). In: *Advances in Comparative and Physiology Biochemistry, Vol. 1,* New York, pp. 217–273.
Nicol, J.A.C. (1963). *Endeavour* 22(85):37–41.
Nilsson, R. (1964). *Acta Chem. Scand.* 18(2):389–401.
Novikov, I.P. and Yangibaev, M. (1966). In: *Proceedings of the First Scientific Conference of Young Specialists, Institute of Cytology of the Academy of Sciences of the USSR* (in Russian), Leningrad, p. 43.
Novikov, I.P. and Barenboim, G.M. (1967). "Apparatus for the Simultaneous Recording of the Bioluminescence Intensity, Oxygen Uptake, and Growth of Luminous Bacteria with Automatic Recording of the Results." *Materials of the Scientific Conference of the Institute of Cytology of the Academy of Sciences of the USSR* (in Russian), Leningrad, p. 66.
Novikov, I.P. and Belozerskaya, N.A. (1967). "Study of the Influence of Inhibitors of Protein Synthesis on Luminous Bacteria." *Materials of the Scientific Conference of the Institute of Cytology of the Academy of Sciences of the USSR* (in Russian), Leningrad, p. 67.
Nurmukhamedov, R.N., Popova, E.G., and Dokunikhin, N.S. (1960). *Opt. i Spektroskopiya* 9(5):593–600.
O'Dwyer, M.F., El-Bayoumi, M.A., and Strickler, S.V. (1962). *J. Chem. Phys.* 36(5):1395–1396.
Osherovich, A.L., Glukhovskii, V.M., and Shapkov, I.S. (1962). *Pribory i Tekhn. Eksperim.* No. 1. pp. 149–154.
Parker, C.A. (1958). *Nature* 182(4641):1002–1004.
Parker, C.A. and Rees, W.T. (1960). *Analyst* 85(1013):587–600.
Parker, C.A. (1963). In: *Advances in Photochemistry Vol. 1*, New York, pp. 306–383.

Parker, C.A. (1964). In: *Advances in Photochemistry Vol. 2.* New York, pp. 306–383.

Perel'man, S.M., Malpnin, V.I., and Giyutte, N.M. (1959). *Pribory i Tekhn. Eksperim.* No. 6, pp. 95–97.

Perelygin, V.V. and Tarusov, B.V. (1966). "Flash of Ultraweak Emission during Damage of Living Tissues." *Biofizika* 11(3):539–541.

Perlman, G.F. (1964). *Biopolym. Symp. 1,* New York pp. 383–387.

Pertsev, A.N., Pisarevskii, A.N., and Soshin, L.D. (1963). *Pribory i Tekhn. Eksperim. No. 5,* pp. 173–176.

Pertsev, A.N., Pisarevskii, A.N., and Soshin, L.D. (1964). *Pribory i Tekhn. Eksperim.* No. 3, pp. 132–135.

Pertsev, A.N., Pisarevskii, A.N., and Soshin, L.D. (1965). *Zh. Prikl. Spektroskopii* 2(5):396–401.

Pertsev, A.N. and Pisarevskii, A.N. (1967). "Measurements Using the Monoelectronic Component of the Noise of Photoelectronic Multipliers." *Pribory i Tekhn. Eksperim.* No. 3. pp. 147–150.

Pesce, A., Bodenheimer, E., Norland, K., and Fasman, G. (1964). *J. Am. Chem. Soc.* 86(24):5669–5675.

Petrusevich, Yu.M. and Ivanov I.I. (1965). *Biofizika* 10(4):698–699.

Petrusevich, Yu.M. and Konoplyannikov, A.G. (1965). *Biofizika* 10(3):524–525.

Petrusevich, Yu.M., Ivanov, I.I., and Tarusov, B.N. (1965). In: *Physical-Chemical Self-Regulation in Cells* (in Russian), Moscow, p. 17.

Pikulik, L.G., Kostko, M.Ya., Konev, S.V., and Chernitskii, E.A. (1966). "Duration of the Excited State of Protein Fluorescence." *Dokl. Akad. Nauk Belorussk. SSR* 10:16.

Pilipovich, V.A. and Sveshnikov, B.Ya. (1958). *Opt. i Spektroskopiya* 4(1):120–166.

Pil'shchik, E.M. and Nikolaeva, M.V. (1963). *Dokl. Akad. Nauk SSSR* 148(1):199–201.

Pil'shchik, E.M., Kudryavtseva, M.V., and Rozanov, Yu.M. (1964). In: *Electron and Fluorescent Microscopy of Cells* (in Russian), Moscow, pp. 98–107.

Pil'shchik, E.M. and Kudryavtseva, M.V. (1965). *Arkh. Anat. Gistol. i Embriol.* 48(4):49–53.

Pisarevskii, A.N., Cherenkevich, S.N., and Andrianov, V.T. (1966). "Study of the Luminescence Spectra and Quantum Yield of DNA in Solution." *Zh. Prikl. Spektroskopii* 5(5):621–624.

Platzmann, R.L. and Frank, J. (1958). In: *Information Theory in Biology,* London.

Pluchennichak, A. (1967). "Some Peculiarities of the Luminescence of Polycomponent Systems with a Nonuniform Distribution of Absorbing Centers." *Materials of the Scientific Conference of the Institute of Cytology of the Academy of Sciences of the USSR* (in Russian), Leningrad, pp. 71–72.

Poglazov, B.F. (1965). *Structure and Functions of Contracting Albumins* (in Russian), Moscow.

Pollard, E.C., Quild, W.R., Hutchinson, F., and Setlow, R.B. (1955). In: *Progress in Biophysical Chemistry Vol. 5,* London, pp. 72–108.

Polyakov, Ya.S., Rozanov, Yu.M., and Brumberg, E.M. (1966). "Apparatus for Investigating the Phosphorescence of Microspecimens." *Tsitologiya* 8(5):677–681.

Polyakov, Ya.S. (1967). "The Phosphorescence Microscope." *Zh. Prikl. Spektroskopii* 7(4):633–637.

Polyakov, Ya.S. and Yagunov, A.S. (1967). "Investigation of the Phosphorescence of Certain Tissues by the Method of Phosphorescence Microscopy." *Dokl. Akad. Nauk SSSR* 176(1):209–211.

Polyakova, G.N., Popov, A.I., and Fogel', Ya.M. (1965*a*). "Characteristics of Photomultipliers for Measuring Weak Light Fluxes." *Pribory i Tekhn. Eksperim.* No. 5. pp. 198–201

Polyakova, G.N., Popov, A.I., and Fogel', Ya. M. (1965*b*). "Study of the Amplitude Distributions of Pulses at the Output of the Photomultiplier." *Radiotekhn. i Elektron.* No.5. pp. 929–935.

Pomoshchnikova, N.A. and Agroskin, L.S. (1961). *Radiobiologiya* **1**(6):838–842.

Popov, G.A. and Tarusov, B.N. (1963). *Biofizika* **8**(3):317–320.

Popov, G.A. (1964). *Application of Chemiluminescent Methods for the Investigation of Stimulated Oxidizing Reactions in the Biosubstratum, Depending on the Degree of Damage, Abstract* (in Russian), Moscow.

Popov, G.A. and Tarusov, B.N. (1964). *Biofizika* **9**(4):528–529.

Powers, E.L., Welb, R.B., and Ehret, C.F. (1960). *Radiation Res. Suppl.* **2**:94–121.

Pringsheim, P. (1949). *Fluorescence and Phosphorescence,* New York.

Prydz, H. and Froeholm, L.O. (1964). *Acta Chem. Scand.* **18**(2):554–556.

Prydz, S. and Rogeberg, T. (1964). *Phys. Norvegica* **1**(4):227–333.

Pshezhetskii, S.Ya. (1962). *Mechanisms of Radiochemical Reactions* (in Russian), Moscow.

Pullman, B. and Pullman, A. (1963). *Quantum Biochemistry,* New York.

Pyatenko, V.S. and Tarusov, B.N. (1964). *Biofizika* **9**(1):134–135.

Rahn, R.O., Longworth, J.W. Eisinger, J., and Shulman, R.G. (1964). *Proc. Nat. Acad. Sci. USA* **51**(6):1299–1303.

Rao, C.N.R. (1961). *Ultraviolet and Visible Spectroscopy, Chemical Application,* London.

Ratner, E.S. (1960). *Opt. i Spektroskopiya* **9**(1):101–107.

Reid, C. (1957). *Excited States in Chemistry and Biology,* London.

Reinstoff, E. (1960). *Erfahrungsheilkunde* **9**(11):521–528.

Rhodes, W.C. and McElroy, W.D. (1958*a*). *Science* **128**(3318):253–254.

Rhodes, W.C. and McElroy, W.D. (1958*b*). *J. Biol. Chem.* **233**(6):1528–1537.

Rice, S.A. and Doty, P. (1957). *J. Am. Chem. Soc.* **79**:15.

Ril', N. (1948). *Energy Migration* (in Russian), Moscow.

Robinson, G.W. and Frosch, R.P. (1962). *J. Chem. Phys.* **37**(9):1962–1973.

Robinson, G.W. and Frosch, R.P. (1963). *J. Chem, Phys.* **38** (5):1187–1203.

Rodionov, S.F. and Frank, G.M. (1934). *Problems in Photobiology and Light Measurements* (in Russian), Moscow.

Rodionov, S.F. and Osherovich, A.L. (1950). *Dokl. Akad. Nauk SSSR* **74**(3):461–464.

Rogers, P. and McElroy, W.D. (1955). *Proc. Nat. Acad. Sci. USA* **41**(2):67–70.

Rogers, P. and McElroy, W. (1958*a*). *Arch. Biochem. Biophys.* **75**(1):87–105.

Rogers, P. and McElroy, W. (1958*b*). *Arch. Biochem. Biophys.* **75**(1):106–116.

Rosen, P. and Edelman, G.M. (1965). *Rev. Sci. Inst.* **36**(6):809–815.

Rosenheck, K. and Weber, G. (1961). *Biochem. J.* **79**(3):29p.

Roshchupkin, D.I. and Vladimirov, Yu.A. (1964). In: *Biological Action of Ultraviolet Radiation* (in Russian), Moscow, p. 90.

Roshchupkin, D.I. (1965). *Studies of Primary Photochemical Processes in Albumins and Aromatic Aminoacids, Abstract* (in Russian), Moscow.

Roshchupkin, D.I. and Vladimirov, Yu.A. (1965*a*). *Zh. Prikl. Spektroskopii* **3**(3):254–260.

Roshchupkin, D.I. and Vladimirov, Yu.A. (1965*b*). *Biofizika* **10**(1):48–55.

Rozanov, Yu.M., Yangibaev, M., and Barenboim, G.M. (1968). "Operation of the Photoelectronic Multiplier with Modulation of the Voltage on Its Dynode." *Pribory i Tekhn. Eksperim.* (in press).

Rozanova, L.M., Barskii, I.Ya., and Brumberg, E.M. (1962). In: *Eighth International Cancer Conference Vol. 4.* (in Russian), Moscow, pp. 68–70.

Rozanov, L.M., Barskii, I.Ya., and Brumberg, E.M. (1963). *Dokl. Akad. Nauk SSSR* **150**(4):907–908.

Rozman, I.M. and Kilin, S.F. (1959) *Usp. Fiz. Nauk* **69**(3):459–482.

Rusinov, L.A. (1962). *Pribory i Tekhn. Eksperim.* No. 3. pp. 141–147.
Russel. G. (1957). *J. Am. Chem. Soc.* **79**(14):3871–3877.
Rybac, B., Lochet, R., and Rousset, A. (1955). *Compt. Rend.* **241**(19):1278–1280.
Samylov, S.V., Tsukerman, V.A., and Model', I.Sh. (1958). *Zh. Eksperim. i Teor. Fiz.* **34**(3):599–607.
Sapezhinskii, I.I. and Emanuel', N.M. (1960). *Dokl. Akad. Nauk SSSR* **131**(5):1168–1170.
Sapezhinskii I.I., Silaev, Yu.V., and Emanuel', N.M. (1963). *Dokl. Akad. Nauk SSSR* **151**(3):584–586.
Sapezhinskii, I.I., Silaev, Yu.V., and Dontsova, E.G. (1964*a*). In: *Free-Radical Processes in Biology* (in Russian), Moscow, p. 47.
Sapezhinskii, I.I., Silaev, Yu.V., and Emanuel', N.M. (1964*b*). *Dokl. Akad. Nauk SSSR* **159**(6):1378–1380.
Sapezhinskii, I.I. and Emanuel', N.M. (1965). In: *Bioluminescence* (in Russian), Moscow, pp. 122–124.
Sapezhinskii, I.I. and Silaev, Yu.V. (1965*a*). *Dokl. Akad. Nauk SSSR* **162**(3):691–693.
Sapezhinskii, I.I. and Silaev, Yu.V. (1965*b*). In: *Controlled Biosynthesis and Biophysics of Populations* (in Russian), Krasnoyarsk, pp. 143–144.
Sapezhinskii, I.I., Silaev, Yu.V., and Dontsova, E.G. (1965*a*). *Biofizika* **10**(3):429–432.
Sapezhinskii, I.I., Silaev, Yu.V., and Emanuel', N.M. (1965*b*). in: *Bioluminescence* (in Russian), Moscow, pp. 102–105.
Schram, E. (1963). *Organic Scintillation Detectors,* Amsterdam.
Seliger, H.H. and McElroy, W.D. (1960). *Arch. Biochem. Biophys.* **88**(1):136–141.
Seliger, H.H. and McElroy, W.D. (1964). *Proc. Nat. Acad. Sci. USA* **52**(1):75–81.
Seliger, H.H. and McElroy, W.D. (1965). *Light: Physical and Biological Action,* New York.
Setlow, R. (1958). *Radiation Res.* **9**(1):179–180.
Setlow, R., Watts, G., and Douglas, C. (1959). In: *Proceedings of the First National Biophysics Congress,* New Haven, pp. 174–183.
Sevchenko, A.N., Konev, S.V., and Katibnikov, M.A. (1963*a*). *Dokl. Akad. Nauk SSSR* **153**(4):875–877.
Sevchenko, A.N., Solov'ev, K.N., Shkirman, S.F., and Sarzhevskaya, M.V. (1963*b*). *Dokl. Akad. Nauk SSSR* **153**(6):1391–1394.
Shchepot'eva, E.S., Ardashnikov, S.N., Lur'e, G.E., and Rakhmanova, T.B. (1959). *Oxygen Effect in the Ionizing Action of Radiation* (in Russian), Moscow.
Shekhtman, Ya.L. (1960). In: *Effect of Peroxides and Oxygen at the Initial Stages of the Radiobiological Action* (in Russian), Moscow, pp. 116–135.
Shelkov, L.S., Prager, I.A., and Kostin, A.G. (1959). *Pribory i Tekhn. Eksperim.* No. 30, pp. 50–52.
Shigorin, D.N., Shcheglova, N.A., and Dokunikhin, N.S. (1962). In: *Physical Problems of Spectroscopy Vol. 1.* (in Russian), Moscow, pp. 292–294.
Shimkivich, V. (1923). *Biological Principles of Zoology, Vol. 1.* Moscow.
Shimomura, O., Goto, T., and Hirata, V. (1957). *Bull. Chem. Soc. Japan* **30**(8):929–933.
Shimomura, O., Johnson, F.H., and Saiga, Y. (1962*a*). *Federation Proc.* **21**(2), Part 2, p. 401.
Shimomura, O., Johnson, F.H., and Saiga, Y. (1962*b*). *J. Cellular Comp. Physiol.* **59**(3):223–239.
Shimomura, O., Johnson, F.H., and Saiga, Y. (1963*a*). *J. Cellular Comp. Physiol.* **62**(1):1–8.
Shimomura, O., Johnson, F.H., and Saiga, Y. (1963*b*). *J. Cellular Comp. Physiol.* **62**(1):9–15.
Shimomura, O., Johnson, F.H., and Saiga, Y. (1963*c*). *Science* **140**(3573):1339–1340.
Shimomura, O. (1964). *Shizen Kagaku Kenkyu Hokoku* **19**(2):47–51.

Shinazo, I. (1964). *Gann* **55**(3):263–266.
Shirokov, V.I. (1956). *Izv. Akad. Nauk SSSR Ser. Fiz.* **20**(3):603–607.
Shishlovskii, A.A. (1961). *Applied Physical Optics* (in Russian), Moscow.
Shlyapintokh, V.Ya. (ed.) (1966). *Free-Radical Processes and Biological Systems,* Moscow.
Shlyapintokh, V.Ya., Karpukhin, O.N., Postnikov, L.M., Zakharov, I.V., Vichutinskii, A.A., and Tsepalov, V.F. (1966). *Chemiluminescence Methods of Investigating Slow Chemical Processes* (in Russian), Moscow.
Shore, V.G. and Pardee, A.B. (1956). *Arch. Biochem. Biophys.* **60**(1):100–107.
Shpol'skii, E.V. and Il'ina, A.A. (1951). *Zh. Eksperim. i Tekhn. Fiz.* **21**(2):142–149.
Shpol'skii, E.V., Il'ina, A.A., and Klimova L.A. (1952). *Dokl. Akad. Nauk SSSR* **87**(6):935–938.
Shpol'skii, E.V. (1959). *Usp. Fiz. Nauk* **68**(1):51–68.
Shpol'skii, E.V. (1960). *Usp. Fiz. Nauk* **71**(2):215–242.
Shpol'skii, E.V. (1962). *Usp. Fiz. Nauk* **77**(2):321–335.
Shpol'skii, E.V. (1963). *Usp. Fiz. Nauk* **80**(2):255–279.
Shpol'skii, E.V. (1965*a*). *Priroda* **10**:22–29.
Shpol'skii, E.V. (1965*b*). *Priroda* **11**:6–16.
Shtompel', V.P., Yushkevich, N.F., Soshin, L.D., and Gribkov, A.I. (1966). "Nature of the Noise of Certain Photoelectronic Multipliers with a Louvered System of Dynodes." *Radiotekhn. i Elektron.* **11**(12):2191–2195.
Shtrankfel'd, I.G. (1963). *Biofizika* **8**(6):690–695.
Shtrankfel'd, I.G. (1964*a*). *Dokl. Akad. Nauk SSSR* **154**(4):953–955.
Shtrankfel'd, I.G. (1964*b*). *Dokl. Akad. Nauk SSSR* **155**(2):461–464.
Shtrankfel'd, I.G. (1964*c*). *Biofizika* **9**(1):68–72.
Shtrankfel'd, I.G. (1964*d*). In: *Electron and Fluorescent Microscopy of Cells* (in Russian), Moscow, pp. 131–136.
Shudel', M.S. (1964*a*). in: *Proceedings of the Fourth Conference on Regeneration and Cell Multiplication* (in Russian), Moscow, pp. 156–157.
Shudel', M.S. (1964*b*). In: *Electron and Fluorescent Microscopy of Cells* (in Russian), Moscow, pp. 88–93.
Shudel', M.S., Chernogryadskaya, N.A., Brumberg, V.A., Razanov, Yu.M., and Brumberg, E.M. (1964). *Dokl. Akad. Nauk SSSR* **157**(2):447–450.
Shudel', M.S. (1965*a*). *Tsitologiya* **7**(3):377–386.
Shudel', M.S. (1965*b*). *Cytophysiological Study of the Mitochondria of Normal and Tumor Cells.* (Candidate's Dissertation, in Russian), Leningrad.
Shudel', M.S., Kudryavtseva, M.V., Pil'shchik, E.M., Pyldvere, V.I., and Bresler, V.M. (1966). "Effects of Certain Aminoazo Compounds on Isolated Liver Cells, Ascites Hepatoma and Ehrlich's Ascites Carcinoma Cells." *Tsitologiya* **8**(4):494–503.
Sidman, J.W. (1958). *Chem. Rev.* **58**(4):689–713.
Siebert, W. and Seffert, H. (1963*a*). *Biochem. Z.* **287**(1–2):92–103.
Siebert, W. and Seffert, H. (1963*b*). *Biochem. Z.* **287**(1–2):104–108.
Sitko, M. (1960). *Sb. Vysoke Skoly Zemedel. Brne* No. 1. pp. 179–181.
Sizov, M.I. (1917). *Living Light* (in Russian), Petrograd.
Skachkov, Yu. F. (1964). *Pribory i Tekhn. Eksperim.* No. 4, pp. 157–159.
Slawinska, D. and Slawinski, S. (1964). *Zeszty Nauk Wyzszej Szkoly Rolniczej Szczecinie* No. 15. pp. 142–145.
Smirnov, V.N., Kullyev, P., Varshavskii, Ya. M., and Spirin, A.S. (1964). *Dokl. Akad. Nauk SSSR* **156**(5):1221–1224.
Soboleva, N.A., Berkovskii, A.G., Chechik, N.O., and Kliseev, R.E. (1965). *Photoelectronic Instruments* (in Russian), Moscow.
Sommermeyer, K., Birkwald, V.B., and Priitz, W. (1961). *Strahlentherapie* **116**(3):354–363.

Spirin, A.S. (1963). *Some Problems in the Molecular Structure of Ribonucleic Acids* (in Russian), Moscow.

Sprince, H., Rowley, G.R., and Jameson, D. (1957). *Science* **125**(3245):442–443.

Stanley, V.A. (1964). *Trans. IEEE PGNS* **11**(3):114–119.

Stauff, J. and Reske, G. (1964). *Naturwiss.* **51**(2):39.

Steele, R.H. and Szent-Györgyi, A. (1955). *Proc. Nat. Acad. Sci. USA* **41**(19):1278–1280.

Steele, R.H. and Szent-Györgyi, A. (1957). *Proc. Nat. Acad. Sci. USA.* **43**(6):477–499.

Steele, R.H. and Szent-Györgyi, A. (1958). *Proc. Nat. Acad. Sci. USA.* **44**(6):540–545.

Steiner, R.F. and Beers, R.F. (1958). *J. Polymer Sci.* **31**(122):53–66.

Steiner, R.F. and Edelhoch, H. (1961). *Nature* **192**(4805):873–874.

Steiner, R.F. and Edelhoch, H. (1962*a*). *Nature* **193**(4813):375–376.

Steiner, R.F. and Edelhoch, H. (1962*b*). *Chem. Rev.* **62**(5):457–483.

Steiner, R.F. and Edelhoch, H. (1963*a*). *Biochim. Biophys. Acta* **66**(3):341–355.

Steiner, R.F. and Edelhoch. H. (1963*b*). *J. Biol. Chem.* **238**(3):931–938.

Steiner, R.F., Lippoldt, E., Edelhoch, H., and Frattali, V. (1964). *Biopolym. Symp. 1*, New York, pp. 355–366.

Stepanov, B.I. and Gribkovskii, V.P. (1963). *Introduction to the Theory of Luminescence* (in Russian), Minsk.

Stepanov, B.N. (1955). *Luminescence of Complex Molecules Vol. 1.* (in Russian), Minsk.

Sternlicht, H., Neiman, G., and Robinson, G. (1963). *J. Chem. Phys.* **38**(6):1326–1335.

Stimson, M.M. and Renter, M.A. (1941). *J. Am. Chem. Soc.* **63**(3):697–699.

Straub, F.B. (1961). *Byokemia*, Budapest.

Strehler, B.L. (1951). *Arch. Biochem. Biophys.* **34**(2):239–248.

Strehler, B.L. and Arnold, W. (1951). *J. Gen. Physiol.* **34**(6):809–820.

Strehler, B.L. and Cormier, M.Y. (1954). *J. Biol. Chem.* **211**:213–225.

Stryer, L. (1960). *Radiation Res. Suppl.* **2**:432–451.

Sturtevant, J.M. (1962). *Biochem. Biophys. Res. Commun.* **8**(4):321–325.

Sukhorukov, B.I., Moshkovskii, Yu.M., Birshtein, T.M., and Lystsov, V.N. (1963). *Biofizika* **8**(3):294–300.

Sushchinskii, M.M. (1950). *Zh. Eksperim. i. Teor. Fiz.* **20**(4):304–317.

Sveshnikov, B.Ya. (1941). *Priroda* No. 2. pp. 15–30.

Sveshnikov, B.Ya. (1952). *Usp. Fiz. Nauk* **46**(3):331–347.

Szent-Györgyi, A. (1957). *Bioenergetics*, New York.

Szent-Györgyi, A. (1960). *Introduction to a Submolecular Biology*, New York.

Tarasov, N.I. (1956). *The Luminescence of the Seas* (in Russian), Moscow.

Tarusov, B.N. (1955). *Biological Effects of Radioactivity* (in Russian), Moscow.

Tarusov, B.N. (1960). *Principles of Biophysics and Biophysical Chemistry* (in Russian), Moscow, p. 1.

Tarusov, B.N., Polivoda, A.I., and Zhuravlev, A.I. (1961*a*). *Biofizika* **6**(4):490–491.

Tarusov, B.N., Polivoda, A.I., and Zhuravlev, A.I., (1961*b*). *Radiobiologiya* **1**(1):150–151.

Tarusov, B.N., Polivoda, A.I., Zhuravlev, A.I., and Sekamova, E.N. (1962). *Tsitologiya* **4**(6):696–699.

Tarusov, B.N. and Zhuravlev, A.I. (1965). In: *Bioluminescence* (in Russian), Moscow, pp. 125–132.

Tarusov, B.N., Kozlov, Yu. P., Urtile, S., and Chzhou Yun-tszen. (1965). *Dokl. Akad. Nauk SSSR* **163**(3):752–753.

Tarusov, B.N., Ivanov, P.I., and Petrusevich, Yu. M. (1967). *Ultraweak Luminescence of Biological Systems* (in Russian), Moscow.

Teale, F.W.J. and Weber, G. (1957). *Biochem. J.* **65**(3):476–482.

Teale, F.W.J. and Weber, G. (1959). *Biochem. J.* **72**(2):15p.

Teale, F.W.J. (1960). *Biochem. J.* **76**(2):381–388.

Teale, F.W.J. (1961). *Photoelec. Spectrometry Group Bull.* **13**:346–348.
Terenin, A.N. (1947). *Photochemistry of Dyes and Related Organic Materials* (in Russian), Moscow.
Terenin, A.N. and Ermolaev, V.L. (1952). *Dokl. Akad. Nauk SSSR* **85**(3):547–550.
Terenin, A.N. (1967). *Photonics of Dye Molecules* (in Russian), Leningrad.
Terpstra, W. (1962). *Biochim. Biophys. Acta.* **60**(3):580–590.
Thorne, C.Y.R. and Kaplan, N.O. (1963). *J. Biol. Chem.* **238**(5):1861–1868.
Tikhonenko, T.I. (1965). In: *Biosynthesis of Protein and Nucleic Acids* (in Russian), Moscow, pp. 193–276.
Tkhor, L.F. and Kozolv, Yu.P. (1965). *Biofizika* **10**(3):523–524.
Toporets, A.S. (1955). *Monochromators* (in Russian), Moscow.
Troitskii, N.A., Konev, S.V., and Katibnikov, M.A. (1961). *Biofizika* **6**(2):238–239.
Troshin, A.S. and Barenboim, G.M. (1963). *Vestn. Akad. Nauk SSSR* No.6, pp. 73–76.
Tsubio, M., Kyogoku, Y., and Shimanouchi, T. (1962). *Biochim. Biophys. Acta* **55**:1–12.
Tumerman, L.A. (1962). *Izd. Akad. Nauk SSSR Ser. Fiz.* **26**(1):84–93.
Tumerman, L.A., Zavil'gel'skii, G.B., and Ivanov, V.I. (1962). *Biofizika* **7**(2):21.
Tumerman, L.A. and Fedorovich, I.B. (1965). In: *Controlled Biosynthesis and Biophysics of Populations, Abstracts* (in Russian), Krasnoyarsk, p. 149.
Turoverov, K.K., Barenboim, G.M., and Sokolenko, V.A. (1967*a*). "A Fluorescence Method of Investigating Conformational Transformations of Proteins Caused by a Temperature Change." *Materials of the Scientific Conference of the Institute of Cytology of the Academy of Sciences of the USSR* (in Russian), Leningrad, pp. 85–86.
Turoverov, K.K., Sokolenko, V.A., and Barenboim, G.M. (1967*b*). "Temperature Quenching of the Fluorescence of Indole Derivatives and Tryptophan Residues of Proteins." *Materials of the Scientific Conference of the Institute of Cytology of the Academy of Sciences of the USSR* (in Russian), Leningrad, pp. 86–87.
Udenfriend, S. (1962). *Fluorescence Assay in Biology and Medicine*, New York.
Udenfriend, S. and Zaltzman, P. (1962). *Ann. Biochem.* **3**(1):49–59.
Udenfriend, S., Zaltzman-Nirenberg, P., and Cantoni, G.L. (1963). *Ann. Biochem.* **5**(3):258–260.
Ungar, G. and Romano, D.V. (1962). *J. Gen. Physiol.* **46**(2):267–275.
Urbach, R. (1930). *Sitzber. Osterr. Acad. Wiss. Math. Naturw. Kl.*, Abt. IIb, **139**:353–362.
Val'dman, M.M. and Sheremet'ev, G.D. (1963). *Izv. Akad. Nauk Ser. Fiz.* **27**(5):693–695.
Van der Ziel, A. (1958). *Fluctuation in Radio-Engineering and Physics* (in Russian translation), Moscow.
Van Duuren, B.L. (1961). *J. Org. Chem.* **26**(8):2954–2960.
Van Es, W.L. and Wisse, J.H. (1963). *Anal. Biochem.* **6**(2):135–143.
Vasil'ev, R.F., Karpukhin, O.N., and Shlyapintokh, V.Ya. (1959). *Dokl. Akad. Nauk SSSR* **125**(1):106–109.
Vasil'ev, R.F., Karpukhin, O.N., and Shlyapintokh, V.Ya. (1961). *Zh. Fiz. Khim.* **35**(2):461–462.
Vasil'ev, R.F., Zhuchkova T.N., and Petukhov, S.M. (1963). *Opt. Prom.* No. 3. pp. 2–3.
Vasil'ev, R.F. (1964*a*) *Luminescence during Chemical Reactions in Solution* (in Russian), Leningrad.
Vasil'ev, R.F. (1964*b*). *Priroda* **12**:22–30.
Vasil'ev, R.F. (1965*a*). *Opt. i Spektroskopiya* **18**(2):236–244.
Vasil'ev, R.F. (1965*b*). *Opt. i Spektroskopiya* **18**(3):415–421.
Vasil'ev, R.F. (1965*c*). In: *Bioluminescence* (in Russian), Moscow, pp. 198–202.
Vasil'ev, R.F. (1966). *Usp. Fiz. Nauk* **89**(3):409–436.
Vasil'ev, R.F. and Chumakova, R.I. (eds.) (1967). *Bioenergetics and Biological Spectro-*

photometry (in Russian), Moscow.
Vavilov, S.I. (1936). *Collected Papers Vol. 1.* (in Russian), Moscow, pp. 438–448. (published 1954).
Vavilov, S.I. (1960). *Microstructure of Light* (in Russian), Moscow.
Vendeneev, V.I., Gurvich, L.V., Kondrat'ev, V.N., Medvedev, V.A., and Frankevich, E.L. (1962). *Bond Energies, Ionization Potentials, and Electron Affinities* (in Russian), Moscow.
Venkstern, T.V. and Baev, A.A. (1965). *Absorption Spectra of Minor Bases, their Nucleosides, Nucleotides, and Certain Oligoribonucleotides* (in Russian), Moscow.
Veselovskii, V.A., Sekamova, E.N., and Tarusov, B.N. (1963). *Biofizika* **8**(1):125–127.
Vil'dgrube, G.S. and Ronkin, Zh.M. (1962). *Pribory i Tekhn. Eksperim.* No. 2, pp. 119–120.
Vil'dgrube, G.S., Dalinenko, N.K., Dunaevskaya, N.V., and Ronkin, Zh.M. (1963). *Pribory i Tekhn. Eksperim.* No. 5. pp. 167–172.
Vil'dgrube, G.S., Ronkin, Zh.M., and Kolosov, Yu.A. (1964). *Izv. Akad. Nauk SSSR Ser. Fiz.* **28**(2):377–387.
Vil'dgrube, G.S., Dunaevskaya, N.V., Manuilov, I.D., Zavarina, M.V., and Vanina, A.V. (1967). "New Series-Produced Photoelectronic Multipliers of the Louvered Type." *Pribory i Tekhn. Eksperim.* No. 2. pp. 159–163.
Vladimirov, Yu.A. and Konev, S.V. (1957). *Biofizika* **2**(1):3–19.
Vladimirov, Yu.A. and Konev, S.V. (1959). *Biofizika* **4**(5):533–540.
Vladimirov, Yu.A. and Litvin, F.F. (1959). *Biofizika* **3**(5):606–609.
Vladimirov, Yu.A. and Litvin, F.F. (1959*b*). *Biofizika* **4**(5):601–604.
Vladimirov, Yu.A. and Burshtein, E.A. (1960). *Biofizika* **5**(4):385–392.
Vladimirov, Yu.A. (1961). *Dokl. Akad. Nauk SSSR* **136**(4):960–963.
Vladimirov, Yu.A. and Chin-ho, Li. (1962). *Biofizika* **7**(3):270–280.
Vladimirov, Yu.A., Litvin, F.F., and Man'-tsi, Tan' (1962). *Biofizika* **7**(6):675–682.
Vladimirov, Yu.A. and Litvin, F.F. (1964). "Photobiology and Spectroscopic Methods." *Handbook on General Biophysics Vol. 8.* (in Russian), Moscow.
Vladimirov, Yu.A. and L'vova, O.F. (1964), *Biofizika* **9**(4):506–507.
Vladimirov, Yu.A. (1965*a*). *Photochem. Photobiol.* **4**(3):369–384.
Vladimirov, Yu.A. (1965*b*). *Photochemistry and Luminescence of Proteins* (in Russian), Moscow.
Vladimirov, Yu.A. and Litvin, F.F. (1965). In: *Bioluminescence,* Moscow, pp. 51–59.
Vladimirov, Yu.A. and L'vova, O.F. (1965*a*). In: *Controlled Biosynthesis and Biophysics of Populations* (in Russian), Krasnoyarsk, p. 150.
Valdimirov, Yu.A. and L'vova, O.F. (1965*b*). "Study of Ultraweak Luminescence of Liver Homogenates and Pulps." In: *Biophysics of the Cell* (in Russian), Moscow, pp. 74–83.
Vladimirov, Yu.A. and Zimina, G.M. (1965). "Luminescence of Certain Proteins and Tryptophan during Monochromatic Excitation in Solutions with Various pH." *Biokhimiya* **30**(6):1105.
Vladimirov, Yu.A., Aksentsev, S.L., and Olenev, V.I. (1965). *Biofizika* **10**(4):614–618.
Vladimirov, Yu.A. (1966). *Ultraweak Luminescence in Biochemical Reactions* (in Russian), Moscow.
Vladimirov, Yu.A., Aike Go, and Ropshchupkin, D.A. (1966*a*). "Investigation of the Primary Photochemical Processes in Proteins. IV. Luminescence Spectra of Primary and Final Photoproducts of Aromatic Amino Acids and Proteins." *Biofizika* **11**(2):237–244.
Vladimirov, Yu.A., L'vova, O.F., and Cheremisina, Z.P. (1966*b*). "Ultraweak Luminescence of the Mitochondria and Its Relationship to the Enzymatic Oxidation of Lipids." *Biokhimiya* **31**(3):507.

Vladimirov, Yu.A. (1967). *Primary Photochemical Reactions in Aromatic Amino Acids and Proteins.* (Doctoral Dissertation in Russian), Moscow.
Voct, D., Gratzer, W.B., Cox, R.A., and Doty, P. (1963). *Biopolymers* **1**(3):193–208.
Volina, V.F. and Lomonosov, I.I. (1966). "Noise and Stability of Photoelectronic Multipliers." *Radiotekhn. i Elektron.* **11**(6):1145–1147.
Vol'kenshtein, M.V. (1965). *Structure and Physical Properties of Molecules* (in Russian), Moscow.
Vol'kenshtein, M.V. (1964). In: *Molecular Biology Problem and Perspectives* (in Russian), Moscow, pp. 172–181.
Vol'kenshtein, M.V. (1965). *Molecules and Life: Introduction to Molecular Biophysics* (in Russian), Moscow. [English translation: Plenum Press, New York (1969)].
Wada, A. and Ueno, V. (1964). *Biopolym. Symp. 1.* New York, pp. 343–353.
Walaas, E. (1963). *Acta Chem. Scand.* 17(2):461–463.
Walton, R.W. (1964). *Rev. Sci. Instr.* **35**(4):518–519.
Wang, S.F., Kawahara, F.S., and Talalay, P. (1963). *J. Biol. Chem.* **238**(2):576–586.
Watanabe, S., Evenson, L., and Culz, V. (1963). *J. Biol. Chem.* **238**(1):324–327.
Weber, G. (1957). *Nature* **180**(4599):1409.
Weber, G. (1960*a*). *Biochem. J.* **75**(2):335–345.
Weber, G. (1960*b*). *Biochem. J.* **75**(2):345–352.
Weber, G. (1961*a*). in: *Light and Life,* Baltimore, pp. 82–107.
Weber, G. (1961*b*). *Nature* **190**(4770):27–29.
Weber, G. (1961*c*). *Biochem. J.* **79**(3):29p–30p.
Weber, G. and Wold, F. (1963). *Federation Proc.* **22**(2): Part 1, p. 348.
Weber, G. and Rosenheck, K. (1964). *Biopolym. Symp. 1,* New York, pp. 333–341.
Weinberg, C.J., Nelson, D.R., Cazter, J.G., and Augenstein, L.G. (1962). *J. Chem. Phys.* **36**(11):2869–2878.
West, W. (1956). *Chemical Applications of Spectroscopy,* New York.
Westermark, T. and Grapendiesser, B. (1960). *Nature* **188**(4748):395–396.
Wetlaufer, D.B. (1962). *Advan. Protein Chem.* **17**:303–390.
White, A. (1959). *Biochem. J.* **71**(2):217–220.
White, E.H., McCapra, F., Field, G.E., and McElroy, W.D. (1961). *J. Am. Chem. Soc.* **83**(10):2402–2403.
White, E.H., Worther, H., Field, G.E., and McElroy, W.D. (1965). *J. Organ. Chem.* **30**(7):2344–2347.
Wilkinson, F. (1964). In: *Advances in Photochemistry,* New York.
Williams, N.T. and Essex, H. (1949). *J. Chem. Phys.* **17**(10):995–997.
Witting, L.A. (1965). *Federation Proc.* **24**(4):912–916.
Wyman, G. (ed.) (1966). "Symposium on Chemiluminescence." *Photochem Photobiol.* 4:6.
Yagunov, A.S. and Prokof'ev, A.Ya. (1966). *Tsitologiya* 8(4):569–572.
Yakushin, V.V. and Tel'nov Yu.Ya. (1963). *Pribory i Tekhn. Eksperim.* No. 4. pp. 120–122.
Yakushin, V.V. (1967). "Monoelectronic Sensitivity of the Photoelectronic Multiplier." *Pribory i Tekhn. Eksperim.* No. 1. pp. 154–157.
Yamazaki, F., Fujiki, K., and Muzata, V. (1965). *Bull. Chem. Soc. Japan* **38**(1):8–12.
Yanoshi, L. (1955). *Zh. Eksperim. Teor. Fiz.* **28**(6):679–694.
Yeargers, E. and Augenstein, L. (1965). *Biophys. J.* **5**(5):687–696.
Zakaryan, A.I. and Tarusov, B.N. (1966). "Inhibition of the Chemiluminescence of Blood Plasma during Malignant Growth." *Biofizika* **11**(5):919–921.
Zakharov, I.V. and Shlyapintokh, V.Ya. (1961). *Kinetika i Kataliz* **2**(2):165–171.
Zalkind, S.Ya. (1937*a*). *Usp. Sovrem. Biol.* **7**(1):50–66.
Zalkind, S.Ya. (1937*b*). *Usp. Sovrem. Biol.* **7**(2):216–233.

Zalkind, S.Ya. (1964). *Byull. Mosk. Obshchestva Ispytatelei Prirody, Otd. Biol.* **69**(6):145–146.

Zhinzhina, G.P., Kruglyakova, G.E., and Emanuel', N.M. (1965). In: *Controlled Biosynthesis and Biophysics of Populations* (in Russian), Krasnoyarsk, p. 144.

Zhuravlev, A.I. (1959). *Effect of Radiation of Lipids and the Lipid Model of the Prophylactic Effect* (in Russian), Moscow.

Zhuravlev, A.I., Polivoda, A.I., and Tarusov, B.N. (1961). *Radiobiologiya* **1**(3):321–324.

Zhuravlev, A.I. and Veselovskii, V.A. (1963). *Bioluminescence* (in Russian), Moscow.

Zhuravlev, A.I., Filippov, Yu.N., and Simonov, V.V. (1964). *Biofizika* **9**(6):671–677.

Zhuravlev, A.I. (1965*a*). In: *Bioluminescence* (in Russian), Moscow, pp. 184–193.

Zhuravlev, A.I. (1965*b*). *Zh. Obshch. Biol.* **26**(2):129–137.

Zhuravlev, A.I., Filippov, Yu.N., and Simonov, V.V. (1965*a*). In: *Bioluminescence* (in Russian), Moscow, pp. 75–89.

Zhuravlev, A.I., Filippov, Yu.N. and Simonov, V.V. (1965*b*). *Biofizika* **10**(2):246–251.

Zhuravlev, A.I., Veselovskii, V.A., and Koshcheenko, N.N. (1965*c*). In: *Bioluminescence* (in Russian), Moscow. pp. 19–50.

Zhuravlev, A.I. (ed) (1966). *Free-Radical Processes in Biological System* (in Russian), Moscow.

Zimmer, K.G. (1961). *Studies on Quantitative Radiation Biology*, Edinburgh.

Zimmerman, H. and Joop, N. (1961). *Z. Elektrochem.* **65**(1):61–65.

Zommermayer, K., Birkwald, V.B., and Prutz, W. (1961). *Strahlentherapie* **116**(3):354–363.

INDEX

S

T

U

V

W

X

GPSR Compliance

The European Union's (EU) General Product Safety Regulation (GPSR) is a set of rules that requires consumer products to be safe and our obligations to ensure this.

If you have any concerns about our products, you can contact us on ProductSafety@springernature.com

In case Publisher is established outside the EU, the EU authorized representative is:

Springer Nature Customer Service Center GmbH
Europaplatz 3
69115 Heidelberg, Germany

Batch number: 10704707

Printed by Printforce, the Netherlands